W9-CXC-170

Technological Innovation
in
Electric Power Generation
1950-1970

Technological Innovation
in
Electric Power Generation
1950-1970

BRUCE A. SMITH

1977
MSU Public Utility Papers

Division of Research
Graduate School of Business Administration
Michigan State University
East Lansing, Michigan

The Institute of Public Utilities, Graduate School of Business Administration, Michigan State University, publishes books, monographs, and occasional papers as part of its program of promoting academic interest in the study of public utilities and regulation. The views and opinions expressed in these works are solely those of the authors, and acceptance for publication does not constitute endorsement by the Institute, its member companies, or Michigan State University.

ISBN: 0-87744-139-1
Library of Congress Catalog Card Number: 76-620054

To Joan and Sparky

Contents

List of Tables

List of Figures

Acknowledgments

This examination of technological advance in the electric power industry has benefited from the careful tutelage of many persons. I am in great debt to Professors Samuel Loescher, George Wilson, James Suelflow, and Ira Horowitz for their advice and constructive criticism of this study throughout its preparation.

My thanks also go to the Office of Economics, Federal Power Commission, which provided computational support for the quantitative analysis in chapter V. Finally, the worth of the unselfish contributions of my colleague Joan Fagerstrom is incalculable. Without her critical inquiries and editorial acumen present at every stage of the study's development, this end product could not have been as refined and comprehensive as it now is.

The opinions expressed in this monograph are solely my own and do not necessarily represent those of the National Science Foundation or other organizations with which I have been affiliated.

I

Introduction

The process of technological advance has received growing public attention as the amount of national resources devoted to technological activity has increased. During the past two decades national research and development (R&D) expenditures have expanded 450 percent, and the number of research scientists and engineers has increased over 250 percent.[1] The data in Table 1.1 show that all sectors of the economy have greatly augmented their R&D expenditures since 1953. The largest change has occurred in the colleges and universities, which have increased their R&D expenditures by over 1,000 percent.

Although the federal government directly undertakes about 15 percent of the nation's total R&D, its role in the process of technological advance has broadened significantly over the past two decades. By the end of the 1960s, two-thirds of the nation's R&D programs were sponsored and funded by the federal government. As technology becomes more complex and costly, further expansion of the government's role is likely.

Private industry, through its own resources and those of the government, now spends over $25 billion on R&D. The industrial sector's growth of R&D expenditures illustrates its increased reliance on technology to realize productive efficiencies in response to both rising production costs and growing demand.

Because of its importance, economists and other social scientists have been prompted to analyze those factors that may influence and accelerate the multifaceted process of technological advance. This process is composed of R&D, invention, innovation, technological diffusion, and the resultant increase in factor productivity.

1

Table 1.1. *National R&D Expenditures by Sector, 1953–1976 (in billions)*

Sector	1953	1961	1969	1976°
Private industry	$3.6	$10.9	$18.3	$26.5
	(71)	(76)	(71)	(70)
Federal government	1.0	1.9	3.5	5.6
	(20)	(13)	(14)	(15)
Colleges and universities	.4	1.2	2.9	4.7
	(07)	(08)	(11)	(12)
Nonprofit institutions	.1	.4	.9	1.2
	(02)	(03)	(04)	(03)
Total	$5.1	$14.4	$25.6	$38.0

SOURCE: National Science Foundation, *National Patterns of R&D Resources: Funds and Manpower in the United States, 1953–1976,* table B-1.
NOTE: Figures in parentheses indicate percentage of column total.
° Preliminary figures.

Even before the days of Abbot P. Usher's research into the causes of invention,[2] economists recognized the important influence of technological advance on economic performance. But only recently have they attempted empirically to document factors that may affect the process.[3]

The most comprehensive examinations of technological advance to date have been conducted by Edwin Mansfield.[4] His research covers many of the components of the process: industrial R&D, technological innovation, and diffusion of innovations. In his econometric studies Mansfield used the number of inventions and innovations produced by the firm to measure technological output, and R&D expenditures to account for technological input. Mansfield was among the first to use the number of publicly disclosed innovations introduced by the firm as a measure of technological performance. This has enabled him to study *innovative* (as distinguished from *inventive*) activity, which plays a central role in the successful commercial utilization of invention to increase factor productivity.

Drawing upon the general methodology utilized by Mansfield, the research undertaken in this study furnishes additional empirical examination of the process of technological advance within a microeconomic framework through an investigation of the technological performance of the electric utility industry from 1950 to 1970. It documents and analyzes this process in an effort to determine the possible influence of the economic environment facing the electric utilities on the industry's technological performance, as measured by technological innovation.

With a view toward producing reliable, accurate, and high-quality primary data about the technological performance of the electric utilities, this analysis examines the influential factors both descriptively and quantitatively. The two approaches complement one another: not all factors can be meaningfully translated into numbers, while some factors are more easily described and interpreted quantitatively.

The electric utility industry is one of the largest industries in the United States and one of the economy's important primary industries, as electric power serves as a fundamental input for much of the manufacturing and service sectors. As regional monopolies, electric utilities are subject to public control—a factor of interest as it may affect technological performance. Additionally, due to the industry's regulated nature, much of its performance data is made publicly available, which greatly facilitates investigation.

In the discussion that follows the *electric utility industry* is defined as the publicly and privately owned companies generating, transmitting, and/or distributing electric power for public consumption; the *electric power industry* refers to the electric utility industry plus the manufacturers of electric power equipment and affiliated trade and research associations.

This investigation deals only with innovation in the fossil-fueled and nuclear (thermal) generation of electric power from 1950 to 1970. Not included in the analysis are innovations in hydroelectric power generation, which has declined to less than 10 percent of total electric power production and is likely to fall steadily during the coming decades, since few economically feasible hydroelectric sites remain undeveloped. The electric utilities have given far greater attention to developments in fossil-fueled and nuclear power generation advances.

Also, technological innovation, as it is defined herein, does not include increases in the size of generating units. Interviews with electric power industry personnel led to the conclusion that increases in generating unit size, per se, usually incorporate extrapolations of existing technology rather than pioneering advances in generating techniques. One of the most widely known technological advances to occur in steam-electric power generation since 1950 was the first commercial use of supercritical steam. The first supercritical unit, a 125 Mw installation, was less than one-half as large as the largest unit placed in service in 1957. Because the analysis in part seeks to examine the relationship between firm size and innovative output, the statistical problem of sampling bias would occur unless the measure of innovative output is size-independent. In order not to penalize unduly either the smaller or the larger systems in measuring their innovative performance, the innovations to be considered must be as size-independent as possible. The largest generating units, however, can be added only to the largest utility systems. For these reasons, the size of a generating unit was not considered an innovation per se.[5]

The process of technological advance is clearly dichotomized in the electric power industry as it is in the railroad, steel, and aluminum industries studied by others. The innovating firm rarely, if ever, is also the inventor.[6] The electric utilities for the most part continue to purchase technology from other firms rather than produce it themselves. Just as farmers purchase technological advances in the form of new seed, fertilizer, or equipment from the manufacturers of such products, the electric utilities buy technology (usually in the form of new capital equipment) from electrical equipment manufacturers, paying for this new technology in the purchase price of the equipment. The manufacturers innovate by pioneering the development and production of power equipment, while the utilities innovate by being the first to install, operate, and apply it to generate electric power.

Although the process of technological advance is composed of R&D, invention, and both types of innovation, this study will examine only the latter type of innovation: the first installation and commercial operation of a new process or piece of equipment to generate electric power.

Data documenting innovations in electric power generation were collected from trade journals and engineering society papers most relevant to the electric power industry. Although all technological advances may not be announced in any single trade or engineering periodical, few significant technological innovations fail to be publicly recorded in at least one of the journals surveyed.[7]

The electric utilities' relative openness in disclosure of technological innovation is prompted by their status as regional monopolies in the distribution of electric power. Although different electric utilities may and do operate generating plants in the same geographic area, a company cannot distribute power to customers in another utility's designated service area. Publicizing their technological advances may benefit the utilities, since potentially significant competition exists between them to attract industrial customers into their service areas, and since many innovations have reduced operating costs and improved service reliability. In addition, many utilities are motivated to announce publicly their technical contributions because of the prestige accompanying the successful innovator.

From the journals surveyed a list of technological innovations was compiled. After the list was formed the utilities that pioneered each development were identified. To ensure that significant innovations were not omitted, the list was screened by individuals familiar with the operations of the electric utilities, including members of the Federal Power Commission, the editorial staffs of industry trade journals, the academic community, and operating utilities, who weighted the innovations according to their technological significance.[8] Weighting of innovations is based on the realistic assumption that all innovations are not equally significant.

In the remainder of the book chapter II describes the structure, composition, and recent performance of the electric utility industry from 1950 through 1970. Particular attention is given to trends in electric power generation and the effects of the industry's structure on technological activity. Chapter III examines in detail the process of technological innovation in the electric power industry, documenting the role of each of the participants in the process and the motives prompting them to engage in innovative activity. Chapter IV identifies the record of technological innovation in fossil-fueled and nuclear power generation by the utilities from 1950 to 1970 and gives a description and history of the most significant. Chapter V presents a quantitative investigation of the relationship between selected characteristics of the innovating utilities and their innovative output, and between electric utilities that have recorded R&D expenditures and their firm size and profitability. In chapter VI the principal findings of the investigation are summarized, some tentative conclusions are offered, and areas where further explanation could provide useful results are identified.

II

The Electric Utility Industry

This chapter analyzes the structure, composition, and recent performance of the electric utility industry from 1950 through 1970. Changes
in the industry's structure and size and in the generating plants' size
and efficiency that have influenced or been affected by technological
innovation are examined in detail. Technological advances have made
it feasible to build larger units, allowing the utilities to expand the
production of electric energy to meet the public's growing demand.
Over the period studied, realization of greater efficiencies in the generation of electric power has enabled the electric companies to sell
larger amounts of power at relatively stable prices despite rising
costs.[1]

Structure and Recent Performance

Industry Size and Growth

Over the past two decades the electric utility industry, which is
among the largest of all U.S. industries, has been as much a construction industry as a seller of energy. It accounts for 10 percent of all
yearly capital expenditures invested by American business; its average
spending for new construction is the highest of any U.S. industry, approximately 10 percent of all new construction for all business,[2] to-
taling over $9.9 billion in 1970.[3] The industry employed over 384,900
persons in that year.[4]

Electric utilities are highly capital intensive: in 1970 the amount of
investment (gross electric utility plant) per electrical department employee was $243,762;[5] $4.71 of capital investment was required to pro-

duce $1 of revenue.[6] The industry's capital-intensive nature is further demonstrated by statistics which show that in 1962 it was among the nation's largest with combined gross capital assets of $69 billion, more than 60 percent greater than the petroleum refining industry;[7] by 1970 industry total assets had climbed to over $99 billion.[8]

The electric utility industry also ranks among the largest domestic industries in terms of sales. Table 2.1, using 1967 data, shows it ranking third in industries where sales data were available.

Table 2.1. *1967 Sales in Selected Industries (in millions of dollars)*

Industry	Sales
Motor vehicles and automotive equipment	31,300
Petroleum and petroleum products	24,822
Total electric utility industry	17,233
Chemicals and allied products	17,067
Industrial machinery and equipment	13,539

SOURCES: *1967 Census of Business*, v. IV, Wholesale Trade Area Statistics, Table 1; and *EEI Pocketbook of Electric Utility Industry Statistics*, 17th ed., 1971, p. 18.

Besides being one of the largest in the nation, the electric utility industry has been one of the fastest growing. Its annual growth in output is about twice the rate of increase in overall industrial production.[9] From 1950 to 1970 total output (kwh sales) for the industry increased by 400 percent, and operating revenues have increased by 330 percent.[10] In comparison, between 1950 and 1970 the nation's gross national product increased by 250 percent.[11] Total electric utility industry sales and revenue data are provided in Table 2.2.

Table 2.2. *Total Electric Utility Industry Sales and Revenues, 1950–1970*

Year	Total energy sales to ultimate customers (billions of kwh)	Total revenue from ultimate customers ($ millions)
1950	280.5	5,086
1955	480.9 (71.5)	8,020 (56.6)
1960	683.2 (42.0)	11,516 (43.5)
1965	953.4 (40.0)	15,158 (31.6)
1970	1,391.4 (46.0)	22,066 (31.4)

SOURCE: *EEI Pocketbook of Electric Utility Industry Statistics*, 17th ed., 1971, pp. 17 and 18.
NOTE: Figures in parentheses indicate percentage increase from prior period.

Growth of productivity also has been rapid. John Kendrick found that from 1899 to 1953 the annual rate of change in total factor productivity for electric utilities increased by 5.5 percent,[12] which was more than three times the annual change in the private domestic economy and the highest yearly change in productivity of any of the thirty-four industries examined.

Since 1953 the trend in factor productivity identified earlier by Kendrick has diminished. Total factor productivity increased by 55 percent from 1953 to 1970.[13] Almost all of this increase is attributed to increases in the capital/labor ratio. The amount of capital used by the privately owned utilities has increased by 250 percent, while total labor employment has increased by 14 percent since 1953. The average yearly change in total factor productivity from 1953 to 1970 has been 2.6 percent, less than one-half the yearly change determined by Kendrick.

Table 2.3. *Number of Electric Utility Systems by Ownership Sector, 1927–1968*

Year	Nonfederal government	REA cooperatives	Federal government	Privately owned	Total
1927	2,198 (50.7)	—	1 (.0)	2,135 (49.3)	4,334
1937	1,878 (54.0)	192 (5.5)	3 (.1)	1,401 (40.4)	3,474
1947	2,108 (54.7)	887 (23.0)	4 (.1)	858 (22.3)	3,856
1957	1,890 (55.8)	1,026 (30.3)	5 (.1)	465 (13.7)	3,386
1968	2,075 (60.2)	960 (27.9)	5 (.1)	405 (11.8)	3,445

SOURCE: FPC, *National Power Survey,* 1970, pt 1, table 2.1.
NOTE: Figures in parentheses indicate the percentage of the row total. May not add to 100 percent due to rounding.

Industry Ownership Sectors

Since the first electric generating plant, New York's Pearl Street Station, began commercial operation over ninety years ago, the electric utility industry has undergone continuous development. The data presented in Table 2.3 illustrate changes in electric utility ownership patterns since 1927. Table 2.4 compares the size of the public and private utilities constituting the electric utility industry using 1970 data.

The largest decline in systems by ownership between 1957 and 1968—a 13 percent reduction—occurred in the privately owned sector. The next largest changes in the number of electric systems between 1957 and 1968 were the nonfederal government utilities, a 9.8 percent increase; and the REA cooperatives, a 6.4 percent decrease.

Privately owned systems. The privately owned electric utilities dominate the entire utility industry despite their few numbers. While

Table 2.4. *Comparative Size Data, Privately and Publicly Owned Electric Utilities, 1970*

	Private*	Public†
Total sales (billions kwh)	1,289.5	395.1
	(76.5)	
Total net generation	1,186.1	346.7
(billions kwh)	(77.4)	
Total generating capacity	262.7	77.6
(millions kw)	(77.0)	
Total electric operating	19.8	3.1
revenue ($ millions)	(86.0)	
Total customers (millions)	56.6	7.4
	(88.5)	

Sources: FPC, *Statistics of Privately Owned Electric Utilities in the United States,* 1970; FPC, *Statistics of Publicly Owned Electric Utilities in the United States,* 1970; and *EEI Pocketbook of Electric Utility Industry Statistics,* 17th ed., 1971.
Note: Figures in parentheses indicate percentage of total electric utility industry, private plus public utility figures.
  * Privately owned Class A & B utilities.
  † Includes federal projects, municipals, and the Power Authority of the State of New York–Niagara and St. Lawrence projects.

representing only 11.8 percent of the nation's electric utility systems in 1968, they account for the largest share of industry sales, revenues, generated output, and customers. The growth of the private sector, like that of the whole industry, has been impressive; since 1950 the privately owned electric utilities' total power generation has increased by 350 percent, an annual growth rate of 6.6 percent.

Among the privately owned systems there is great disparity in system size. A large number of private firms are small companies serving nonurban, nonindustrial markets. Like the small publicly owned utilities, these firms are usually distribution-only systems, purchasing their power requirements from the larger utilities.

Within the privately owned sector the large utilities account for a disproportionate share of generated output, assets, and revenues. These large systems have grown in size both through internal expansion and through acquisitions of other electric utilities. In 1970 the fifty largest privately owned electric utility systems, composed of ninety-three separate operating companies, generated 64 percent of the industry's net kwh output. These fifty systems represent less than 2 percent of the nation's operating electric systems. The 212 Class A & B utilities,[14] representing 51 percent of the privately owned operating utilities in 1968, accounted for more than "98 percent of the privately owned electric light and power industry assets and revenues."[15] By

1970, the then 210 Class A & B utilities accounted for nearly 100 percent of the private sector's assets and revenues.

These relatively few systems thus exert a pervasive influence on the entire industry's economic and, as will be demonstrated later, technological performance. Much of the analysis of the utilities' performance in the following chapters will focus on these privately owned utilities.

Nonfederal government systems. The largest number of electric utility systems, the nonfederal government systems, include city and town municipal utilities, county and state systems, and special utility districts. In 1970 these systems generated about 9 percent of industry production.[16] In 1923 there were 3,084 such systems, but by 1968 the number had declined from this peak to 2,075 with 1,369 of them purchasing all of their power requirements from either privately owned or federal systems.[17] The municipal systems are the most numerous of the nonfederal systems and vary greatly in size. Some serve only a few hundred customers, while others like the Los Angeles Department of Water and Power serve over a million.

REA cooperatives. The Rural Electrification Administration (REA), founded in 1936, has promoted the increased use of electric energy in rural America through the creation of rural electric service cooperatives. Federal financing of such cooperatives was necessary due to the apparent reluctance of private industry to provide electric service in rural areas. This reluctance was caused in large part by the relatively high distribution costs per customer, making rural electrification less profitable than urban service areas. The REA systems serve an average load density of about four customers per mile of line.[18]

When the cooperatives were first organized in the 1930s, they were almost exclusively distribution systems. As time passed and their loads grew, generation and transmission cooperatives were developed to supply the cooperatives' power requirements. The REA cooperatives now purchase 77 percent of their wholesale power requirements; in 1940, 92 percent was purchased. Their largest single power source is the government sector, including federal systems, which supplied 45 percent of the cooperatives' requirements; the privately owned sector supplied 32 percent.[19]

Federal systems. The federal systems account for the second largest segment of industry capacity and generated energy.[20] The five largest federal systems include the Bonneville Power Administration, the Southwestern Power Administration, the Southeastern Power Administration, the Department of the Interior's Bureau of Reclamation, and the nation's largest system, the Tennessee Valley Authority (TVA). Unlike the four other federal systems, TVA now operates fossil-fueled and nuclear generating capacity in addition to hydroelectric facilities. These five systems sell most of their electric power to other publicly owned systems (municipals and cooperatives) in their operating regions, to a small number of large industrial purchasers, and to government agencies such as the Energy Research and Development Administration for nuclear diffusion and processing operations.

Operating Costs

Labor, capital, and fuel costs are the major operating costs faced by the electric utilities and have been increasing steadily. From 1958, when data were first available, to 1970, total salaries and wages for reporting Class A & B privately owned utilities increased 185 percent to $2.71 billion; the total employment of privately owned utilities over this period grew from 348,300 to 384,900 in 1970.[21] While this wage increase is sizeable, when one learns that industry output, measured in total kwh generated, rose 240 percent, labor cost is put in perspective. Also reducing the relative impact of labor cost is the utilities' increased investment per employee. Such investment has risen by 330 percent since 1950, to $243,762 per employee in 1970.[22]

In an industry where the capital/labor ratio is high and where more than four dollars of capital investment is required on average to produce one dollar of revenue, the cost of capital becomes a prime economic consideration. The Handy-Whitman Public Utility Construction Index measures changes in the cost of constructing an electric power plant and in the cost of its components. Table 2.5 shows that each portion of the index more than doubled between 1950 and 1970, with the largest jump occurring in the last five years. In the same period total construction expenditures for production, transmission, and distribution of power by the privately owned electric utilities increased by 400 percent.[23]

Fuel cost is the single largest production expense faced by an electric utility over the expected lifetime of a plant, representing between 70 and 80 percent of total production costs. An average-sized fossil-fueled plant may burn more than one million tons of coal in one year. Larger plants can consume up to three million tons of coal, up to seven million barrels of oil, or up to seventy billion cubic feet of gas per year.[24] Table 2.6 illustrates the increase in the utilities' consumption of

Table 2.5. *Price Indexes for Selected Components of Construction, 1950–1970*

Handy-Whitman Public Utility Construction Index	1950	1955	1960	1965	1970
Building°	67	83	104	111	142
Electric light and power†	66	84	102	107	135

SOURCE: *Statistical Abstract of the United States,* 1971, table 1094, p. 664.
NOTE: Based on data covering public utility construction costs for 95 items in 6 geographic regions. Covers skilled and common labor; does not reflect tax payments nor employee benefit costs. (1957–59 = 100.)
° Includes cost of components for power plant building construction.
† Includes cost of material and equipment for steam-electric plant generation (boilers, turbine-generators, coal and ash handling equipment, condensors and tubing, and cranes); includes separate listing for operations employees.

fuels and their real cost. The dollar cost per ton of coal equivalent fuel[25] increased by 24 percent from 1950 to 1970. The real fuel cost in cents per million BTU, by contrast, increased by 115 percent over the same period. The greatest change occurred during the period between 1965 and 1970. Together, the increased fuel costs and rising consumption of electricity raised fuel expenditures of privately owned electric utilities from $.76 billion in 1950 to $3.73 billion in 1970, almost a 400 percent increase.[26]

Table 2.6. *Total Electric Utility Industry Fuel Consumption and Cost, 1950–1970*

	1950	1955	1960	1965	1970
Millions of tons of coal equivalent fuels	138	207	266	368	573
Price per ton of coal equivalent fuel°	$5.95	6.27	6.62	6.23	7.39
Real fuel cost (cents per million BTU)†	21.6	20.7	25.9	24.3	47.0

SOURCE: *EEI Pocketbook of Electric Utility Industry Statistics,* 17th ed., 1971, pp. 22 and 23.
° Data refer only to privately owned electric utilities.
† Deflated by wholesale fuel price index.

These increased labor, capital, and fuel costs have motivated the utilities to employ generation techniques that: (1) automate power generation to reduce labor cost and increase efficiency and reliability; (2) realize scale economies by substituting capital in place of labor and fuel; and (3) raise thermal efficiencies to lower fuel costs. Technological advances such as these help hold down the sale price of electricity despite increased operating costs. These advances illustrate the utilities' desire to minimize the unit labor and capital costs and the amount of fuel required per unit of output.

Trends in Power Generation

As the demand for electric power has continued to grow, the electric utilities have increased their generating capacity. New and larger fossil-fueled and nuclear base load generating units account for the vast majority of the 330 percent increase in industry generating capacity between 1950 and 1970.

The data in Table 2.7 illustrate the industry's growing reliance on nuclear and fossil-fueled generating capacity to meet the rising demand for energy. Earlier predictions of much larger nuclear generating capacity have not been met, due, in part, to environmental problems and siting considerations. Because of these factors, it is estimated that an electric utility should allow nearly ten years' time from its initial consideration of building a nuclear unit to the unit's commercial operation. By contrast, the lead time for nuclear power units built in the early and late 1960s was five and seven years, respectively. When it is remembered that over this period the industry talked of the demand for power doubling almost every decade (the figures in Table 2.2 appear to corroborate this statement), one can appreciate the costs of such development and construction time. Despite these growing lags the utilities remain committed to the development of nuclear capacity, as "approximately 50 percent of the expenditures currently being made for the various research and development programs by the electric utilities and their trade associations is going into the nuclear power field."[27] This evidence suggests that the electric utility industry has set a high priority on meeting much of the projected increase in electric energy sales with nuclear capacity additions.

The growth of other methods of power generation also has been deterred for several reasons. Pollution abatement techniques have yet to be perfected, slowing down current fossil-fueled capacity additions. Most practical hydroelectric sites already have been developed. Pumped storage facilities[28] have revitalized some of these sites; however, location of additional hydroelectric plants is becoming a serious problem. Internal combustion engine facilities, often using diesel-powered and gas turbine units, are used mostly for short-term, peak load operation and are relied on to supply base load power only during emergencies.

Table 2.7. *Net Generation of Electricity, Class A & B Utilities, 1950–1970*

Type	1950	1955	1960	1965	1970
Fossil-	214,944	365,511	517,397*	735,601	1,077,450
fueled	(80.0)	(87.7)	(89.6)	(91.1)	(90.7)
Nuclear				3,725	19,113
Hydro:				(0.5)	(1.6)
conventional	53,164	50,662	59,622	67,042	71,436
	(19.8)	(12.2)	(10.3)	(8.3)	(6.0)
pumped					3,423
storage					(0.3)
Internal	660	521	362	707	16,067
combustion	(0.2)	(0.1)	(0.1)	(0.1)	(1.4)
Total	268,768	416,694	577,381	807,075	1,187,489

SOURCE: FPC, *Statistics of Privately Owned Electric Utilities in the United States,* years indicated.
NOTE: Figures in parentheses represent percentage of column total; exclude station use.
* Nuclear generation included in fossil-fueled generation.

Plant size and efficiency. Electric utilities have met the continued increase in demand for electric power over the twenty-year period in large part by building bigger units. From Table 2.8, twenty-three fewer plants accounted for almost five times as much generating capacity in 1970 as in 1950. This table shows the change in the number of plants of specific capacities, evidence of the growing dependence on larger units by the electric utilities. By 1970, 7.1 percent of all reporting plants had capacities greater than 1,000 Mw. Plants over 100 Mw capacity represented two-thirds of all reporting plants in 1970, compared to only one-quarter of all reporting plants in 1950.

Other FPC data available since 1956 (presented in Table 2.9) document the increase in average plant size. Between 1956 and 1970 installed capacity increased by more than 180 percent, and the average plant size increased almost 200 percent. Industry personnel often mention this increase in average plant size as a significant development in the electric utility industry. The major reason for the increase is the utilities' interest in realization of the economies of scale in the generation of electric power.[29] Electric utilities at times must choose between investing their capital in plants using new, unproven technology or in plants of larger size using more conventional technology. (The larger-sized units can employ new technology, but this technology usually is founded on extrapolation of existing unit sizes rather than on departures from the prevailing generating techniques.) De-

Table 2.8. *Class A & B Privately Owned Electric Utilities Steam Plant Capacity, 1950–1970*

Year	Total steam plants	Over 100 Mw cap.	Over 500 Mw cap.	Over 1,000 Mw cap.	Total Mw cap.
1950	684	141	—	—	44,633
1951	689	163	—	—	49,797
1952	683	170	—	—	53,488
1953	695	196	—	—	60,013
1954	703	220	11	—	67,957
1955	691	241	—	—	75,406
1956	685	258	—	—	79,462
1957	693	269	—	—	85,693
1958	693	302	30	—	99,896*
1959	698	324	39	—	110,717
1960	711	346	47	—	120,124
1961	701	359	59	9	128,175
1962	693	402	67	10	136,762
1963	681	378	72	12	143,146
1964	671	387	83	14	151,215
1965	653	398	89	17	159,141†
1966	651	408	95	19	167,106
1967	657	418	112	26	180,726
1968	657	426	127	36	196,787
1969	656	432	140	43	209,950
1970	661	442	155	47	220,536

SOURCE: FPC, *Statistics of Privately Owned Electric Utilities in the United States,* years indicated.
— Data not available.
* Reflects restatement of capacity in existing stations due to a change in reporting requirements.
† After 1965 steam plant capacity does not include nuclear plant capacity.

pending on the utility's policies and cost considerations, either one of these choices may be made.

In the generation of electric power one facet of economic efficiency is productive efficiency. From an engineering viewpoint this productive efficiency can be measured by the heat rate and thermal efficiency of the plant. The more efficient the plant, the lower is the heat rate measured by units of generated heat (BTU) per net unit of output (kwh).

Thermal efficiency measures how effectively available inputs, including labor, plant, fuel(s), and cooling facilities, are employed to produce electric power. The higher the thermal efficiency, the more effectively these inputs are being used. To examine the changes in productive efficiency two measures are used: (1) the national average

Table 2.9. *Average Steam-Electric Plant Size in the Total Electric Utility Indus-try, 1950–1970*

Year	Number of plants	Installed capacity (Mw)	Average plant size (Mw)
1956	1037	92,591	89
1957	1039	99,500	96
1958	1051	110,633	105
1959	1051	122,982	117
1960	1060	132,818	125
1961	1050	142,226	135
1962	1056	150,677	143
1963	1058	166,216	157
1964	1061	175,575	165
1965	1052	188,213	179
1966	1070	198,341	185
1967	978	210,798	216
1968	979	216,020	231
1969	978	241,439	246
1970	979	258,040	264

SOURCE: FPC, *Steam-Electric Plant Construction Cost and Annual Production Expenses,* years indicated.

NOTE: Includes no nuclear, geothermal, or gas turbine generation plants; only fossil-fueled steam-electric generating plants.

heat rates for fossil-fueled steam-electric plants in the total electric power industry; and (2) the number of steam-electric plants operating under specific heat rates of 10,000, 9,500, and 9,000 BTU/net kwh.

Figure 2.1 shows that the number of selected electric plants operating under the above heat rates has risen greatly since 1950. The number of plants operating under 10,000 BTU/net kwh rose from two in 1950 to 121 in 1970. Plants included in Figure 2.1 are the most efficient in the industry.

Figure 2.2 gives more general data about the average heat rate and thermal efficiency for fossil-fueled steam-electric plants in the total electric utility industry. Average heat rates dropped by 25 percent and thermal efficiency rose by 33 percent between 1950–1970. This is a major accomplishment when facing diminishing returns[30] and is due in large part to the engineering skill of the equipment designers and operators. Further significant improvements in thermal efficiency depend on the development of economical methods of raising operating temperatures and pressures. Such improvement will require new technology in the areas of high-temperature physics and metallurgy.

Figure 2.1. *Selected Plant Heat Rates, 1950–1970*

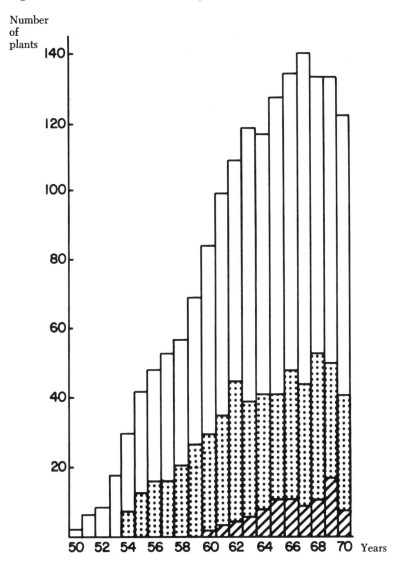

SOURCE: FPC, *Steam-Electric Plant Construction Cost and Annual Production Expenses,* years indicated.
☐—plants operating under 10,000 BTU/net kwh.
▫—plants operating under 9,500 BTU/net kwh.
▨—plants operating under 9,000 BTU/net kwh.

Figure 2.2. *National Average Heat Rate and Thermal Efficiency for Fossil-Fueled Steam-Electric Power Plants, 1950–1970*

Heat
rate*

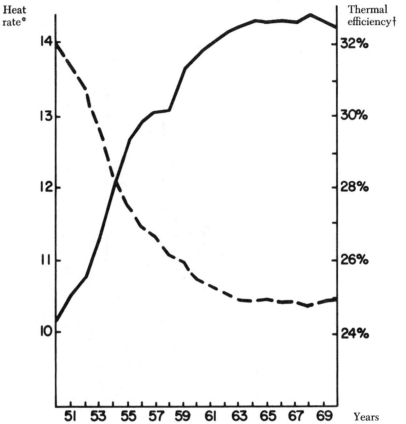

Thermal
efficiency†

Years

SOURCE: FPC, *Steam-Electric Plant Construction Cost and Annual Production Expenses, 1970,* table 9, p. xxxi.

NOTE: Internal combustion plants included prior to 1958.
  * Expressed in 1,000 BTU/net kwh (- - -).
  † Based on 3,413 BTU as the energy equivalent on one kwh (—).

Industry R&D

The electric utility industry historically has spent little money of its own on research and development (R&D). The Federal Power Commission found in 1970 "an apparent low level of such [R&D] expenditures" by the electric utilities. In that year the electric utilities spent $46,037,000 on R&D[31] (.23 percent of their total electric operating revenues),[32] significantly less than R&D expenditures of other primary industries. For example, the ratio of company R&D funds to net sales for the electrical equipment manufacturers, which supply the utilities with the majority of their equipment, was 3.7 percent in 1969; for the aircraft industry, 4.3 percent; and for the primary metals industry, 0.8 percent. The all-industry average for company R&D funds to net sales in 1969 was approximately 2 percent.[33] Thus, while their output grew at twice the national rate, the electric utilities spent about one-tenth the national average on R&D.

Effects of Industry Structure on Technological Activity

Four elements in the composition of the electric utility industry have affected its technological performance: (1) The industry consists of regional monopolies where competition for customer sales or revenues historically has been limited. (2) Many of the industry's smaller and publicly owned utilities do not generate their own power requirements. (3) The number of pooling and interconnection agreements between utilities has increased rapidly. (4) The existence of public regulation of electric utilities has affected both the level of the utilities' R&D expenditures and the utilities' incentives to perform innovative activity.

Regional Monopolies

As regional monopolies the electric utilities only compete directly for customer sales when new loads, especially industrial concerns, may decide to locate in a firm's service area or on the interface between two service areas. Because such competition is present in a limited way, individual utilities do not face losses in business from making disclosures about their research activities to other electric utilities. According to industry personnel, this lack of direct competition promotes free dissemination of research results throughout the industry.[34]

Not everyone agrees with this. After publication of the Electric Research Council's *Survey of Research*, a compendium of utility industry R&D projects, many of the listed research projects were found to examine similar topics. If the utilities communicated with each other about all of their research activity, they might have avoided unnecessary duplication of research projects that already had been explored and completed.

On balance, while the results of all research do not seem to be freely disseminated, there does seem to be a genuine effort on the part of most utilities to conduct more cooperative research. The need for such research was belatedly recognized by the industry by the mid-1960s, when the individualistic impulses of utility managements, intensified by market regionalism, had been stilled by the growing necessity for increased R&D expenditures.

Distribution-Only and Hydroelectric Utilities

In 1968, 2,419 systems (about 70 percent of the total number of electric systems in the United States) only distributed electric power. These privately owned, municipal, and cooperative systems did not generate their own power requirements; they purchased power from other, usually larger, electric utilities. In addition to these distribution-only utilities are electric systems that generate their power needs primarily from hydroelectric facilities. Every large federal power agency now generates its power requirements by hydroelectric means (only TVA also has steam-electric power plants).

These utilities and federal power agencies probably do not have the incentive to perform research in thermal-electric power generation simply because they would not directly benefit by its results. Instead, what technological activity they do sponsor or conduct is devoted mainly to developing more efficient transmission and distribution systems. The remaining 30 percent of the electric utility systems, usually the larger privately owned systems that generate their base load power requirements by either fossil-fueled or nuclear means, are responsible for what utility R&D is performed in the area of power generation advances.

Pooling and Interconnection Agreements

The first power exchange, the Connecticut Valley Power Exchange, was organized in 1925. Interconnections and pooling agreements, permitting electric utilities to purchase power from other participating companies that have power to sell, have increased both in number and importance as transmission system coordination and load frequency controls were introduced and made commercially feasible, and as system reliability and generation techniques improved.[35] These agreements have been organized for two reasons, each related to the process of technological advance.

(1) Rising demand for electric energy and the resulting load growth faced by the individual utilities necessitated plant capacity additions. Pooling agreements eliminated the need for every member simultaneously to increase plant capacity. The pool structure has allowed staggered construction of new capacity among all the pool members. The members' service reliability is increased, since power can be purchased from another member company when needed during an emergency or an unexpected increase in system demand. The pool structure also reduces the members' required reserves, as all the pool participants share their reserve requirements.

(2) Utilities also have formed pools to realize additional production economies of scale. These organizations often collectivize the means of production and transmission. The pool members' distribution markets, while served by the individual utilities, are interrelated through the collective generation facilities, allowing the units built by and for the pool to be of much larger capacity than is justifiable for an individual utility. The member utilities thus benefit by realizing greater production scale economies.

For example, Philadelphia Electric Company's pooling agreement with Public Service Electric and Gas and Pennsylvania Power & Light allowed the utility to build very large units—some units as large as 22 percent of the total system load.[36] This figure is two to three times greater than the accepted optimal size for individual units operating on a single utility system.[37]

The operation of fewer and larger generating units consequent to the growth of pooling agreements has created important problems with respect to technological advance. The opportunities to innovate and pioneer are diminished as fewer generating units are being built. The increase in unit size escalates the total cost of possible innovation as construction costs rise and more expensive equipment is developed and installed. These historic trends particularly have influenced the process of technological advance in the utility industry over the twenty-year period by limiting the ability of an individual electric utility to undertake power generation innovations.

The industry's technological performance will continue to be affected, since the formation of power pools shows no signs of subsiding. A growing number of electric utilities, wishing to realize anticipated economic and technical advantages accruing to pool members, have proposed the formation of larger and more comprehensive utility groups. It has yet to be decisively proved whether such collective organizations of electric utilities either promote or discourage innovative activity. While the number of generating plants continues to decrease (see Table 2.9), it is not axiomatic that power pools will abstain from incorporating technological developments in new generating units. On the one hand, power pools and their associated members may be more able to finance the rising costs of such developments as the ownership of generating stations can be jointly allocated. On the other hand, pools may be interested only in promoting reliability and scale economies and unwilling to engage in the risky business of generating electric power with new types of equipment designed by the manufacturers. The pool management may view the cost of failure as prohibitively expensive. Whether or not the pools will take advantage of the curtailed opportunities to innovate in power generation depends primarily on how the pool and its members define the organization's objectives.

Since the controlling interests in most power pools are held by the utility industry's largest companies, the continued growth of pools can serve only to increase the technological responsibilities of these utilities. The impact of the largest utilities on the rest of the industry will expand regionally and nationally. Thus the growth of power pools portends not only potentially higher costs for technological advances and curtailment of the opportunities for innovation in power generation, but also most probably alteration of the structure of the technological advance in the utility industry by extending the technological dominance of its largest utilities. Each of these effects will have important, but as of yet, unmeasureable consequences for the electric utility industry and the energy-consuming public.

Regulatory Influence

Although the effect of public regulation of electric utilities on technological advance is not fully understood,[38] partial conclusions can be drawn from information gathered in this study. Appendix A provides a brief description of the history of regulation in the industry.

Industry personnel disagree about regulation's influence on technological activity. William F. Crawford, president of Republic Flow Meters Company, division of Rockwell Manufacturing Company (an equipment supplier of the utilities), states that regulation discourages research by the utilities as it controls utility profits. Because profits are regulated there is less reward for research by producers of electric power.[39] Donald C. Cook, chairman of American Electric Power Company, has stated, however, that "the attitude of most of the regulatory bodies in approving and allowing research and development expenditures has been very constructive, and they are to be highly commended for it."[40] Philip Sporn has said that local, state, and/or federal regulation, contrary to the industry's expectations, have not acted as barriers to the utilities' innovative activity.[41]

In general, it appears that regulation has neither promoted nor induced innovative activity by the electric utilities. This conclusion has been reached in part because regulation historically has not compensated the utilities for their efficient use of productive resources. The regulatory process has neither rewarded those utilities with relatively low costs of service nor penalized those with relatively high costs of service. Few, if any, stimuli to improve the utilities' economic or technological performance have originated within the regulatory framework. In fact, the effect of regulation on the utilities' cost of service generally has been recognized to perpetuate the status quo.[42] Such perpetuation obviously does little to promote financial commitments to technological activity designed to improve the firm's technical performance.

It is possible that a period of regulatory lag, a time of delayed public scrutiny, may provide a financial inducement to engage in innovative activity. If the utilities perceive a period of regulatory lag as an opportunity to earn a supranormal rate of return, they might be willing to finance cost-reducing technological activity with these larger returns in order to realize greater financial gains in the future. The usual response of electric utilities to regulatory lag, however, has not been to perform innovative activity but, rather, to realize increased financial returns in the short run.

Although the regulatory process may not have promoted innovative activity, it may have induced, through the impact of the Averch-Johnson (A-J) effect,[43] a capital-using bias among those power generation innovations that have been initiated. Stated simply, the A-J effect results in the reduction of the regulated utility's shadow price of capital, and therefore defines the firm's allocation of productive resources to be nonoptimal since capital becomes artificially cheaper than other factor inputs. Being able to earn a financial return only on its capital expenditures, the utility will overinvest in rate base (capital) items in order to maximize its returns. Regulation may thus promote expansion of the rate base, but not with capital-saving equipment. The record of innovative activity presented in chapter IV illustrates that relatively few innovations in electric power generation were capital saving.

Finally, while the motive to reduce operating costs through innovation would more likely emanate from the electric utility managements than from the regulatory framework, the great majority of utility executives evidently have been content to forego active development of such cost-reducing technology.

Another facet of regulation's impact on electric utilities' technological activity is its effect on the level of utilities' R&D expenditures. Regulation has not compelled utilities to forego R&D ventures by unduly limiting their financial returns. While it is true, of course, that regulation can constrain the upward and downward fluctuations of the utilities' rate of return, the average return of those firms reporting R&D expenditures was significantly higher than that of all Class A & B privately owned electric utilities. From 1966 to 1970 the average rate of return on their rate base for those Class A & B utilities reporting R&D expenditures to the FPC was 7.56 percent. The average rate of return for all Class A & B privately owned utilities (not just those reporting R&D expenditures) from 1966 to 1970 was 7.35 percent, significantly less (at the .05 confidence limit) than the above figure.

Although the FPC was concerned about the lack of electric utility participation in industry R&D, the commission did not undertake any direct action creating incentives to increase R&D expenditures until 1970. Before that time the FPC adopted a policy characterized by benign neglect. While the state commissions each could formulate policy concerning R&D, few did so. The electric utilities' expressed hope of having a consistent regulatory treatment for R&D went unrealized. Thus until 1970, the major regulatory impact on R&D expenditures of the electric utility industry was one of insufficient moral suasion.

The central problem facing the utilities with regard to R&D and regulation was, and continues to be, to persuade the regulatory commissions to allow R&D expenditures either into the utility's rate base or into its operating expenses. Many state commissions evidently felt that R&D was too risky to be allowed as an operating expense and that the equipment manufacturers were in a better position to undertake such ventures. Utility personnel now think that the state commissions have become more reasonable (lenient) in their allowances for R&D.[44]

Unless R&D expenses could be placed in these accounts, the firm would not be able to earn a return on this expenditure (as a business expense). The utilities also were concerned about the financial consequences of incurring R&D outlays that could prove to be unsuccessful. Without some special provisions, the firms did not know whether they could recoup such losses and were reluctant to undertake many R&D projects.

Prior to 1970 the FPC Uniform System of Accounts, used by fifteen of the state regulatory commissions,[45] limited the recording of R&D expenses to administrative and general expense accounts. There was no attempt to introduce R&D into functional operating accounts. When the FPC realized that moral suasion alone was not inducing the utilities to increase R&D activity, this ambivalence ended. Financial suasion was introduced when the FPC renovated its accounting treatment of utility R&D expenditures.

After submitting proposed accounting changes for R&D expenditures to the industry for comment, on 3 September 1970 the FPC ruled to include utility R&D outlays in two new accounts.[46] The R&D costs that coincide with the development of facilities for utility operations were included in a plant account category, a subaccount of construction work in progress. Utility R&D expenditures not related to a specific project and not related to construction, such as R&D contributions to the EEI Research Division, and those related solely to the utility's operations were classified as deferred debits. The proposal to treat R&D outlays as deferred debits was preferred by accountants under traditional regulatory accounting principles, which attempt to diminish the impact of exceptional outlays through amortization. The introduction of a plant account category gave the utilities more assurance that R&D outlays would receive consideration as a rate base item.

Although the accounting treatment is somewhat different for each type of R&D expense, the R&D expenditures in both accounts may be included in the utility's rate base. As a plant account, the subaccount of work in progress may be handled in one of two ways: (1) the subaccount may be included directly in the utility's rate base; or (2) carrying charges (interest during construction) may be calculated and capitalized (included in the rate base). If the second option is taken, then the plant subaccount itself, not just the accompanying interest charges, will be included in the rate base when the work in progress is completed and transferred into the regular operating (plant) accounts. The principal difference between these choices open to the commissions is not whether the subaccount is put into the rate base, as it will be in either case, but at what time the subaccount is included.

As part of the rate base the utility immediately may begin to recover all its R&D expenses included in the construction work in progress account. Only that portion of R&D costs that remain in the deferred debit account at the end of the accounting period may become a rate base item. The probability that the R&D expense entered in construction work in progress will become a rate base item is almost certain since, when construction is complete, the plant becomes part of the rate base. Previous regulatory consideration indicates, however, that the year-end R&D balance in the deferred debit account is not always allowed as a rate base item, since usually it is not part of plant or equipment. Those balances not allowed in the rate base by the state regulatory commission may be amortized over a five-year period as an operating expense.

Most of the utility's R&D outlays in the deferred debit account of a minor or of a recurring nature are transferred to the appropriate operating expense account. A portion of those deferred debit R&D expenditures that are large and nonrecurring and cause a distortion in yearly company expenses may be amortized to the appropriate operating expense account. The utility also may transfer large, unsuccessful R&D outlays to the deferred debit account so that a portion of these outlays may be amortized. Thus, if the electric utility spends R&D funds successfully improving a piece of capital equipment, the firm may realize an interest payment during its construction with these expenditures entering the rate base after completion. If the utility spends R&D funds on an unsuccessful piece of equipment, it may reduce the financial burden of the loss through amortization.

On 30 April 1973 the FPC amended its 1970 R&D accounting and reporting procedures to "further encourage industry's research commitments." The new decision provided that companies can come to the commission for advance rate base assurances for R&D expenditures greater than $50,000 undertaken by the utility or as part of a project undertaken by other utilities or organizations. (In 1970, more than one-half of the 219 electric utilities listed by the FPC reported spending less than $50,000 on R&D.) Also, at the company's request the FPC will "track" R&D expenditures of more than $50,000 when recorded under the R&D expenditures account.

These accounting changes, belatedly introduced, seek to promote utility R&D outlays. Under the 1970 accounting revisions, R&D outlays on which the utility does not earn a return as part of the rate base can be fully recovered through charges to operating expenses. The 1973 amendments further ensure that R&D expenditures will be treated as rate base items. While it may be premature to determine if such changes in regulatory accounting have exerted a significant influence on R&D funding by the electric utility industry,[47] they have provided more assurance to the electric utilities of reasonable and consistent rate treatment of R&D costs. Also of note, between 1970 and 1973 the average reported percentage of total electric utility operating revenues devoted to R&D expenditures rose from .23 to .82 percent.

The previous discussion has noted how certain characteristics of the entire electric utility industry may have affected its technological performance. To complement this industry analysis, the characteristics of individual firms in the electric utility industry that may influence a company's technological performance now are identified and examined.

Corporate Strategy and Structure

The electric utilities, relying on the manufacturers of electric equipment to initiate and provide the bulk of technological inventions in electric power generation, have contributed to technological advance in the electric power industry in two ways: (1) by innovating in being the first company to generate power from new equipment supplied by the manufacturers and assuming the associated risks of such pioneering efforts; and (2) by financing and conducting related R&D activities. Although both types of contributions are at times indiscernible, the following discussion concentrates only on the first of the utilities' contributions to technological advance.

Corporate Strategy

Within the electric utility industry the technological leaders are the few large, privately owned utilities that account for the vast majority of industry revenues, assets, and R&D expenditures, since the required investments for adopting new technology can exceed the total financial resources of all but the largest utilities. However, not all large utilities exercise technological leadership. The number of research-oriented firms is small and, for the most part, unchanging. In an industry that contained 3,445 electric systems in 1968, only thirty-three of them have exerted an innovative impact on electric power generation in the period since 1950.

Ultimately, the factor causing some utilities to perform innovative activity and others not to do so is the utility's corporate strategy. The majority of electric utilities' managements willingly have allowed the manufacturers and the few research-oriented utilities to dominate the industry's technological record. This strategy ensures that most utilities have defined their technological role as followers and not as leaders. The following examples illustrate the types of strategic decisions made by utilities that directly have influenced their innovative performance.

Scale economies versus new designs. The largest electric utility system in the United States, TVA, has not exercised significant leadership in electric power generation innovation, as defined in this study. Like some other utilities, TVA has chosen to specialize in existing power generation techniques. With few exceptions TVA has followed the policy of building larger units using proven designs rather than building units incorporating new designs (technology). "TVA has not tried to design the most efficient units in the U.S. The high cost of such units make [sic] them uneconomical unless fuel costs are high.... This [is] consistent with TVA's practice of not pioneering in too many technical areas with a single unit. It is believed that with its large system, TVA can contribute best to the art of steam generation by developing large units."[48] The management of TVA has instituted a policy whereby a conscious effort is made to realize further economies of scale using more or less conventional technology. Since it was concluded that increases in size, per se, are not technological innovations for this study, TVA's policy of favoring scale economies versus new technology has lessened its technological impact on the industry. To TVA and most other electric utilities, savings from scale economies are greater than savings from introducing new generation advances.[49] Such a view fails to recognize that scale economies are finite and cannot always be gained beyond a certain productive capacity.

Construction time and plant design. Because demand has grown rapidly, the utilities have had to expand their capacity. Given the long lead times required for power plant construction, in some cases the electric utility can meet future power demands only by building new capacity as quickly as possible. Instead of economies of scale versus new technology, the choice here is between building capacity additions as quickly as possible to be available by a specific date, or experimenting with new generation designs. In many cases the utility has chosen expediency, building plants incorporating conventional technology with assurance of meeting the delivery date.[50] The case of Philadelphia Electric's Peach Bottom II reactor illustrates this situation.

The Peach Bottom I reactor was a high-temperature, gas-cooled type which incorporated a recognized innovative design. The second reactor, using more conventional technology, was a boiling water reactor whose basic design had been previously used by other utilities and approved by the AEC. As Philadelphia Electric needed this new capacity as soon as possible, it employed the proven design that the company felt was more easily implemented having been successfully used before.

Another example of a utility's choice of conventional designs versus innovative ones is the design of Niagara Mohawk Power Corporation's Nine Mile Point Nuclear Power Station. Niagara Mohawk realized the possible costs of exploring an unproven design and did not want to be too innovative. According to then vice-president and chief engineer Minot H. Pratt, "we want as few [technical] changes and innovations as possible. We are interested in not having much unusual about. It will be as much as possible a carbon copy, in this size, of past designs."[51]

This policy, like that of Philadelphia Electric, was introduced to minimize the problems and delays of adding new capacity to the company's system. It is, of course, legitimate to want to minimize avoidable delays. Successful technological advance, however, often requires continual inquiry by many different organizations. Generalization about the electric utility industry's innovative behavior is tenuous at best when made from only two observations; but utilities should recognize that a policy of constructing plants as fast as possible, which seems to exclude the use of untested designs, may include too much technological complacency by assuming that new designs will be tested by other companies.

Corporate Structure

The structure of the firm also may influence its innovative activity. According to company officials, the decision to separate research work from other business activity has enhanced research within the firm by creating an atmosphere conducive to technological experimentation. Two technologically oriented firms, American Electric Power and Detroit Edison, have separate divisions where design innovations can originate. American Electric Power has created much of its engineering-design orientation and developed its record of technological performance through its American Electric Power Service Company. The AEP Service Company is responsible for the equipment and plant design work and research activity of its parent company. Detroit Edison has conducted some of the required R&D for the Fermi fast breeder reactor project as well as other research projects by organizing a research division within the firm. Other electric utilities, generally the larger utilities, perform R&D within the firm using such research divisions. Overall, however, few electric utilities have chosen to create separate research staffs within their corporate structure.

Corporate strategy may thus affect innovative activity of the electric utilities in several important ways: by deciding to realize scale economies rather than improving the efficiency of the fuel cycle; by choosing conventional plant and equipment designs rather than new, untested ones to avoid possible operational delays; and by organizing the firm's internal research structure.

The following chapter will examine in detail the process of technological advance in the electric power industry and will identify the utilities' motives for performing technological activity.

III

The Process of Technological Advance
in
The Electric Power Industry

The electric utility industry has relied increasingly on technological advance to keep pace with society's growing demand for electric power. This growth of demand spawned in part the energy crisis and has contributed to the environmental crisis. The magnitude of the energy-environmental crisis has been perceived by a growing number of persons inside and outside the electric utility industry.[1] This awareness has forced utilities to reexamine the propriety of certain operating practices and policies, as solutions to these crises may be in conflict with one another. Although no panacea has yet been found, the existing partial solutions have depended on further refinements of existing technology and creation of new ways to produce electric power without rendering the natural environment uninhabitable. This chapter will examine two components of the process of technological advance in the electric power industry: the organizations participating in the process of research and development and the utilities' motives for conducting technological activity.

The analysis is facilitated by the homogeneous nature of electric power. Over the years there have been many refinements in the production process while the final product has remained unchanged. Since changes in product quality of electricity (excluding the increased reliability of electric service) are insignificant, this examination will consider only changes that have occurred in production.

Refinements in the process of generating electric power have been numerous since 1950. Sixty documented technological innovations, presented and examined in chapter IV, range from the first commercial atomic power plant and the use of supercritical steam temperatures and pressures[2] to the conveying of coal by a slurry pipeline from mine to generating station. Industry spokesmen laud the utilities' performance and progress; yet the writer has encountered several utility executives who have wondered why their industry was of any interest. To these executives, at least, the electric utilities merely provided a service that is, more often than not, taken for granted. The existence of this view among some electric utility managements may be due to the fact that, despite many refinements, the fundamental technology of converting fuel energy to electric power has remained unglamorous and unchanged since New York City's Pearl Street Station started operation in 1882.[3] As is the case in most industries, the process of technological advance is an incremental one: the contribution of any one organization may be small, but the interactions of several industry groups produce important improvements in technology.

**Participants in the Process
of R&D and Innovation**

Five types of organizations are responsible in varying degrees for technological R&D and innovation in the electric power industry: (1) the government sector; (2) the manufacturers of power equipment; (3) the electric utilities, alone or in groups;[4] (4) industry associations, such as the Edison Electric Institute (EEI), the Electric Research Council (ERC), and the Electric Power Research Institute (EPRI); and (5) colleges and universities.

The size of R&D expenditures and the type of research required for a project have a direct bearing on the role of these participants in the process of technological advance. Tables 3.1, 3.2, and 3.3 show the allocation of energy research expenditures between private industry and government in FY 1963 and FY 1973. As the research becomes more basic and less applied, the amount of government-supplied funds seems to increase relative to those of industry. Most of the R&D performed in the area of oil and gas research is funded from private industry sources, whereas nuclear R&D expenditures are funded mainly by the government.

Within the electric power industry most basic research has been conducted by the equipment manufacturers. Much of the research fi-

Table 3.1. *Estimated Government and Industry Expenditures for Energy R&D, FY 1963 (in millions)*

Type of research	Federal govt.	Private industry
Coal	$ 11.0	$ 11.0
Petroleum and natural gas	40.0	336.0
Nuclear:		
fission	210.0	90.0
fusion	26.0	3.0
Electricity:		
generation	1.0	94.0
transmission	.3	62.0
Solar	3.0	.5
Magnetohydrodynamics (MHD)*	9.0	4.0
Fuel cell*	8.2	9.0
Thermionics*	11.0	2.3
Thermoelectricity*	11.0	3.0
Total	$330.5	$614.8

SOURCE: *Energy R&D and National Progress,* Energy Study Group, 1964, table 1-25, p. 30.
NOTES: Energy R&D includes research performed by the petroleum, natural gas, and coal industries and their equipment suppliers, as well as that conducted by the electric utilities and manufacturers of power equipment. The figures in this table were compiled from a survey made by the Energy Study Group; industry numbers are order-of-magnitude, and one may include research in another commodity.
* These methods of generating electric power, incorporating fundamental changes in technology, are described in Appendix E.

nanced and performed by individual utilities is "applied research," solving practical, operating problems of their generation, transmission, and distribution systems. Thus the character of a needed research venture often determines which organization undertakes the research. The equipment manufacturers conduct R&D most suited to their resources, such as developing new equipment design. When a research laboratory is required, the manufacturers or universities are more suited to conduct the research. When the project necessitates testing an electric power system's performance, the utilities are usually more qualified to perform the research.

The distinction is important, as the manufacturers apply their R&D to building better-designed, larger, more efficient, and more reliable types of *power equipment,* whereas the utilities' research often applies these individual capital improvements to projects attempting to make their entire *power systems* more efficient and reliable.

According to Table 3.1, the government spent over $330 million in

1963 on "energy R&D" (which includes more than just electric power
R&D), allocating over 70 percent to nuclear fission and fusion
research. Excluding the first two categories of research because they
are not part of the electric power industry, private industry outspent
the government in only three areas: in the generation and transmission of electricity and in research on the fuel cell. The figures in
Tables 3.1 and 3.2 clearly show that government-sponsored and -con-
ducted research has continued to play a very important role in total
power industry R&D.

Table 3.2. *Estimated Government Expenditures for Energy R&D, FY 1973*
      (in millions)

Type of research	Amount	Sponsoring agency	Amount
Coal	$ 94.4	AEC	$435.9
Petroleum and natural gas	26.1	Dept. of Interior	116.5
Nuclear:		EPA	30.5
fission	356.3	NSF	18.1
fusion	65.4	TVA	16.8
Stationary source emission control	52.5	HUD	2.8
Improved housing systems	2.8	Nat'l Bur. Standards	1.0
Energy systems studies	2.2		
Other technology	21.9		
Total	$621.6		$621.6

SOURCE: Office of Science and Technology.

  Tables 3.2 and 3.3 present FY 1973 data for government and private
sector energy R&D expenditures. Although not all of the data are
directly comparable with those in Table 3.1, several observations
about energy R&D since 1963 can be made. While government-financed R&D in petroleum and natural gas dropped 35 percent in the
decade following 1963, government-sponsored coal research increased
to $94.4 million, 850 percent of its 1963 level, making the coal indus-
try the most heavily research-subsidized of the fossil-fuel industries.
Much of this R&D funding has been spent by the Department of the
Interior in coal gasification and desulfurization projects.
  The increasing importance of nuclear power as an energy source is
shown by the size and growth of the government and private sector's
nuclear R&D expenditures. Nuclear-related government R&D expenditures, representing nearly 70 percent of the total government energy

Table 3.3. *Estimated Private Sector Expenditures for Energy R&D, FY 1973 (in millions)*

Industry	Amount
Coal[1]	$ 6.0
Petroleum[2]	500.0
Petroleum equipment manufacturers	50.0
Natural gas[3]	100.0
Natural gas equipment manufacturers[4]	25.0
Electric utilities[5]	44.0
Electric equipment manufacturers[5]	105.0
Nuclear power[6]	300.0
Unspecified manufacturing and nonmanufacturing	20.0
Total	$1150.0

SOURCE: National Science Foundation, Office of Advanced Technology Applications.
NOTE: Figures should be regarded as approximations.
[1] Includes R&D expenditures of support industries.
[2] Includes exploration and production R&D; excludes field testing of equipment ($1.5 billion).
[3] Excludes R&D spent on drilling and capitalization of new transportation and harbor facilities.
[4] Excludes home appliance R&D.
[5] Excludes all nuclear R&D.
[6] Includes R&D of electric utilities and equipment manufacturers.

R&D in both FY 1963 and 1973, increased by nearly 80 percent from 1963, illustrating the government's commitment to a future nuclear-based energy system. By 1973 the AEC's energy R&D expenditures were nearly four times as large as those of any other agency. The private sector also allocated an increasing proportion of its total energy R&D effort to nuclear power, rising from 15 percent in 1963 to 26 percent of R&D expenditures in 1973. The electric utilities and equipment manufacturers spent more than twice as much on nuclear power R&D as on all other types of energy R&D in 1973.

The growth of total energy R&D since 1963 has been impressive; government and private sector energy R&D increased to $1.77 billion in 1973, an 87 percent increase in ten years. Somewhat surprisingly, the proportion of total energy R&D financed by the government remained a constant 54 percent over the ten-year period 1950–1970, although there were shifts in several categories of energy R&D. For example, the ratio of federal to private R&D expenditures in the coal industry increased significantly from 1.0 in 1963 to 15.5 in 1973; while the ratio for nuclear-related R&D dropped from 2.5 in 1963 to 1.4 in

1973, again illustrating the growing priority private industry places on nuclear R&D.

Although not comparable to Table 3.1 on a dollar basis, the data in Table 3.4 illustrate the R&D expenditures of the major sectors in the electric power industry. The relative size of R&D expenditures of these groups remained stable throughout the decade since 1963. In 1973, as in 1963, the equipment manufacturers accounted for the predominant share of power industry R&D expenditures. If we make the realistic assumption that the equipment manufacturers are responsible for at least three-fourths of the nuclear power R&D listed in Table 3.3, then the manufacturers accounted for 73 percent of total power industry R&D in 1973. While this figure increased only slightly since 1963, when nearly 65 percent of total power industry R&D was expended by the equipment manufacturers, the change does indicate that the manufacturers' position in energy-related R&D remained dominant despite the utilities' growing commitment to R&D.

Table 3.4. *R&D Outlays by Group in the Electric Power Industry, 1963*

Group	R&D outlays	Percent
Manufacturers	$ 97,560,000	64.4
Electric utilities	53,180,000	35.1
EEI	757,578	.5
Total	$151,497,578	100.0

SOURCE: *EEI Bulletin* 32 (July 1964): 167.
NOTE: R&D outlays presented herein are not comparable to industry figures presented in Table 3.1 due to dissimilar industry definitions.

The Government Sector

In addition to its regulatory impact discussed in chapter II, the government affects technological advance in the power industry in two primary areas: nuclear power and pollution control.

Nuclear power. Over two-thirds of total energy R&D expenditures by the federal government is spent on nuclear power R&D. Most of these expenditures are funded through the Energy Research and Development Administration, which supplanted the Atomic Energy Commission (AEC), with many projects contracted to private industry. The yearly financial reports of the AEC reveal expenditures of more than $1.5 billion on central station nuclear power reactor R&D through 1971. Table 3.5 documents the growth of these expenditures since 1960, when such figures became available. The main goal of government-sponsored nuclear R&D has been the commercial production of electric power from nuclear steam generators on a competitive basis with fossil-fueled plants. Many of the innovations in nuclear power generation have been funded either directly or indirectly by the AEC's Power Demonstration Reactor Program.

Table 3.5. *Atomic Energy Commission R&D Expenditures on Civilian Nuclear Power Reactors R&D Costs, 1960–1971 (in millions)*

Year	Expenditures	Year	Expenditures
1960	$100.0	1966	$120.0
1961	103.0	1967	140.9
1962	94.3	1968	166.4
1963	88.1	1969	154.4
1964	98.4	1970	147.4
1965	116.5	1971	155.7

Source: *U.S. Atomic Energy Commission Financial Reports,* years indicated.

These innovations are not all patentable, because many times the background research has been performed by or for the government, making it public property. The policy of not allowing the participating companies to have a proprietary interest in the development of government-sponsored R&D programs probably reduces the appeal of taking part in such projects.

The government originally entered nuclear reactor research to develop a power plant for its submarine fleet. Not surprisingly, in the early stages of the civilian power reactor program there was considerable spillover from Navy reactor research developed under Admiral Hyman G. Rickover. In fact, the first so-called commercial atomic power was generated from a land-based submarine reactor in 1955.[5]

Shortly after Admiral Rickover selected Westinghouse Corporation to design and build the submarine reactors, General Electric entered the Navy program. Both companies established their own laboratory facilities and have grown to become the two largest suppliers of nuclear steam supply systems in the United States.

The AEC's Power Demonstration Reactor Program changed character as it developed its research programs. As the newly formed nuclear power industry gained knowledge about nuclear power reactors, the need for detailed research guidance by the AEC diminished. At the program's start the AEC attempted to explore various methods of generating electric power from nuclear energy by encouraging private industry and the utilities to participate. This required building different types of reactors and analyzing their performance to discover which method of producing steam from nuclear fission was most economically and technically feasible.

In this "first round" of the program, industry was invited to participate with the AEC in financing prototype reactors on a name-your-own-terms basis. Projects developed under these conditions were the Yankee Atomic Electric Company's Rowe reactor and the Detroit Edison-Atomic Power Development Associates' Enrico Fermi fast breeder reactor.[6] The first operational nuclear power reactor, the Shippingport Project,[7] was financed and owned by the AEC. Duquesne Light Company bought the steam produced by the reactor and piped it into the utility's turbine-generator to produce electric power.

The terms in the "second round" were altered so that the AEC financed the design and construction of the reactor with the utility (usually a public power organization or cooperative) paying for the turbine-generator facilities, plant site, and buildings. Terms were modified again for the "third round," increasing the utility's share of the project costs. Under these stipulations, the utility (this time most often a privately owned utility) designed, constructed, and operated the entire project. The AEC helped to finance the R&D for the project and subsidized the fuel charges for five years.[8]

While government-sponsored power reactor research has not diminished in size, research programs have moved from emphasizing research in many different areas of nuclear technology performed by various power industry groups to examining a few selected areas of nuclear technology by a few sponsored organizations.

These changes in the direction and type of research conducted through the AEC power reactor programs have resulted in lengthened research project lifetimes. The benefits accruing to the participating utilities are now realized over a longer period of time. For example, the AEC contract was awarded to Duquesne Light Company in 1953 for the nation's first commercial nuclear power station. Electricity was produced from the reactor in 1957. By contrast, the Commonwealth Edison-TVA-AEC fast breeder reactor project funded in part through the AEC was announced in January 1972 and has a target date for first power generation in the 1980s.

The federal government continues to spend large amounts of money in nuclear power R&D. The formation of the Energy Research and Development Administration and the high priority assigned to the development of a commercial fast breeder reactor ensures a significant commitment for R&D expenditures over the coming decade.

Pollution control. Pollution control is a second area where the government has affected innovation in electric power generation, though less directly than in the nuclear field. Local, state, and, more recently, the federal government through ordinance and legislation have forced utilities to change certain operating practices. Electric power industry R&D, spurred by public and government concern, recently has focused its attention more and more on solving pollution problems.

Unlike the nuclear area where government funds have supplied the financial impetus for innovation, in the area of pollution control the various levels of government are demanding that the utilities and other business enterprises identify the sources of pollution and then create or employ technology that will solve the pollution problems.

For example, government-imposed limits on the temperature differential of incoming and outgoing plant cooling water have necessitated new and improved types of cooling facilities. Restrictions on the particulate and gas content of stack effluents have led to more effective filtering equipment.[9] As reported in *Electrical World,* the Orange County (California) Board of Supervisors voted at the end of 1969 to halt additional construction of fossil-fueled plants in the county until the utilities could supply evidence to the Air Resources Board that such power stations could be operated without emitting pollutants into the air.[10] Other government agencies have issued similar, though less drastic, rulings.

The Equipment Manufacturers

The manufacturers of electrical equipment[11] historically have been responsible for much of the development of power equipment and for proceeding with the research necessary to ensure future industry growth. Next to the federal government, the manufacturers have initiated and performed the greatest part of industry R&D. Reasons which account for their dominance are: the competitive and proprietary nature of research among the manufacturers; the limited nature of research benefits accruing to an individual, R&D-sponsoring electric utility; and the level of financial and technical resources necessary for some research projects.

The manufacturers of electrical equipment not only compete with each other for profits from the sales of their products, but also in technical research.[12] The continued rises in the demand for electricity and in production expenses have required newer, larger, and more efficient power equipment developed and produced by the manufacturers. In many cases flexible purchasing policies enable utilities to buy from the manufacturers that produce the best qualified equipment.

This technical competition between manufacturers is illustrated by the following example given by Larry Dwon of the American Electric Power Company (AEP). When a need developed for AEP to use high-voltage, high-speed circuit breakers in their transmission system, the company approached various domestic equipment suppliers to determine if such equipment could be developed and built. The suppliers told AEP that with the present level of technology such circuit breakers could not be economically manufactured. AEP then made inquiries of foreign producers of such equipment.[13] A Swiss firm agreed to develop the circuit breakers and produce them for the AEP system. Not long afterward, the domestic suppliers were marketing similar circuit breakers.

This close relationship between successful equipment marketing and technical ability that has evolved in the electrical equipment industry has no parallel in the electric utility industry proper, where market regionalism promoted by public regulation allows little direct competition for residential sales of electricity.

The prospective profits and patent royalties that the manufacturers hope to realize from the sale of their equipment provide the impetus and means for research. This financial stimulus is not present if and when the utilities perform research projects. Only limited benefits would accrue to an individual utility were it to finance a major research undertaking such as developing a new piece of power equipment. Since a utility's purchases of a given piece of equipment are few in number, the company's savings from such R&D expenditures probably would be small. Whereas the benefits to the individual utility from such a project probably would not be great enough to warrant such expenditures, the manufacturer selling equipment to many utilities can realize a large enough return from the development of a new piece of equipment to justify the undertaking.

The manufacturers are not subject to the regulatory framework that has provided few inducements to the utilities to undertake R&D projects. As mentioned in chapter II, a standardized regulatory accounting treatment of R&D expenditures was not employed until 1970. In addition, since the rates determined to be just and reasonable by regulatory commissions are not usually founded on efficiency criteria, there is little incentive for electric utilities to undertake technological activity to produce factor-saving technology. Regulation in general does not promote innovative activity, because the current rates set by the regulators are based on the utility's past economic record and preserve the status quo rather than promote more efficient resource allocation.

The electric utilities continually have relied on the manufacturers, partly because the research and testing laboratories of the equipment suppliers were already operating, partly because the manufacturers possessed greater financial and technical resources at their disposal. The utility industry has perceived the construction of its own laboratories as an unnecessary expense, duplicating existing facilities of the manufacturers. For these reasons and others to follow, very few utilities have chosen to employ the research staffs needed to conduct even R&D programs that are modest in comparison to those of the manufacturers.

The Electric Utilities

Although much of the industry R&D is performed through the leadership of the equipment manufacturers, the electric utilities make many technical contributions. While the individual utility may be unable to finance or staff research programs on the manufacturer's scale, the utilities conduct research in areas such as field installation of equipment, system operating efficiencies, equipment reliability, load characteristics, distribution economics, and transmission system operation. Both publicly and privately owned utilities face common operating problems that require technical examination through nearly identical research programs. The historic antagonism between public and private power, however, influences the innovative process insofar as there has been only limited coordination of public and private utilities' R&D efforts.

The privately owned electric utilities have assumed the leading role in most utility R&D, coincident with their predominant position in producing electric energy in the United States.[14] A small group of privately owned utilities dominates R&D activity of the private sector and, consequently, of the entire electric utility industry. The fifty largest privately owned systems, which in 1970 generated 64 percent of the industry output, accounted for 90 percent, $41.5 million, of the

1970 R&D expenditures of the Class A & B privately owned firms reported by the FPC.

Table 3.6 breaks down the 1970 R&D expenditures of the Class A & B utilities. The largest portion of R&D performed inside the companies, one-third of total utility R&D, was devoted to generation, with nuclear generation receiving the greatest funding. When the utilities' R&D support to nuclear power groups is added to the nuclear plant R&D, the size of the utilities' commitment to a nuclear-based power system in the coming years is clearly illustrated; 31 percent of these utilities' R&D expenditures are devoted solely to nuclear-related R&D.

Table 3.6. *Class A & B Privately Owned Electric Utility R&D Expenditures, 1970 (in thousands)*

Type of research	Amount	Percent
I. R&D performed inside the firm		
Generation:		
hydroelectric	$ 863	1.9
fossil-fueled steam	6,523	14.2
nuclear	7,383	16.0
internal combustion	545	1.2
direct conversion	26	.1
System planning, engineering, and operation	2,158	4.7
Transmission	2,835	6.1
Distribution	1,754	3.8
Other	3,449	7.5
Subtotal I	$25,536	55.5
II. R&D performed outside the firm		
Research support to		
Edison Electric Institute	$ 4,789	10.4
Nuclear power groups*	7,046	15.3
Others	8,665	18.8
Subtotal II	$20,500	44.5
Total (I + II)	$46,036	100.0

SOURCE: FPC, *Statistics of Privately Owned Electric Utilities in the United States,* 1970, p. 741.

* See Appendix G for a representative sample of nuclear power groups and their member utilities.

One of the problems in measuring R&D contributions by utilities is taxonomic in nature. Work done by utility engineers on projects such as the development of a new cable or substation is not always called R&D, but is classified as engineering. This has been the commonly accepted way of accounting for what could be called research.[15] This method of classification artificially reduces the size of the utilities' R&D expenditures. The accounting procedures of other electric utilities may designate all engineering work as R&D, which has the effect of inflating research expenditures. These unresolved differences in accounting have caused continued public misunderstanding about the utilities' role in power industry R&D.

Examples of research conducted by individual utilities are: (1) the investigation into the influence of atmospheric humidity and the sulfur content of coal on electrostatic precipitator capacity; (2) the search for a commercial utilization for flyash, a waste material produced by burning coal; (3) the examination of the causes of low-order furnace explosions; and (4) the investigation of methods of storing energy so that off-peak electric rates may be applied to the utilities' space-heating and air conditioning load. (If successful methods of storing heat and electric energy can be developed, then this energy can be produced during periods of lower demand, and therefore at lower costs, and then stored to be used later.)[16]

The results of these and similar research projects often are given to the equipment suppliers, allowing the manufacturers to improve the design and performance of their products. Although industry sources disagree about the frequency or effectiveness of the utility-manufacturer communication, it does seem to offer potentially great benefits to the innovative process in the electric power industry. As expected, the engineering design- and research-oriented electric utilities play a larger part in this communication than utilities that follow others' technical leads. Utility-manufacturer communications are an integral part of the innovative process; they provide feedback to the manufacturers about their products' reliability and operation under working conditions.

In 1962 the chairman of the FPC, Joseph C. Swidler, created static in the electric utility industry by charging that R&D activities of privately owned utilities had no organization, no central purpose, and were too small and haphazard.[17] Whether correct or not, Swidler's comments performed a useful service by forcing the electric utilities to become more cognizant of their responsibility toward conducting a share of power industry R&D. With this awareness individual utilities have attempted to make more contributions to continuing the technical advancements necessary for industry growth. The types of projects described above have since been carried out by more utilities. Electric companies also have attempted, although not entirely successfully, to publicize their R&D efforts to give notice to the public (and possibly to the utilities themselves) that the manufacturers do not account for *all* of the R&D in the electric power industry.

Although the number of utilities sponsoring industry research has increased, participation is far from universal. Even with occasional special price inducements from the manufacturers, many utilities remain wary of sponsoring and participating in pilot projects, or of adopting technology that is not yet fully developed commercially. Such technological reticence on the part of utilities has important implications for the path of technological innovation and diffusion in the industry.

In further reaction to Swidler's remarks, groups of utilities began to fund and conduct more joint research projects. By joining in R&D programs the utilities could increase the financial and technical scope of utility-sponsored research and reduce duplicate efforts. Financial reasons, such as better bond distribution and the large research, development, and construction costs, help to encourage such multiutility efforts. As the AEC power reactor program began to emphasize private financing of nuclear projects, the inducement for utilities to create group research efforts increased. This group research activity represents, in addition to that of the individual utility, the major type of R&D performed by the electric utilities. The following examples typify group utility R&D programs:

1. A group of electric utilities and a manufacturer[18] have sponsored a basic study of the principles of magnetohydrodynamics (see Appendix E) to examine its commercial application to electric power generation.

2. Another group of electric utilities has financed basic research with the Gulf General Atomic Corporation in the area of thermonuclear fusion.[19]

3. The largest research effort on the part of the utility groups is in the application of nuclear fission to the generation of electric power. Many of the nuclear power plants operating today have received financial and, at times, technical aid from utility research groups. Often one utility directs the projects with the other members contributing financial and technical resources. Examples of this form of research organization are the Dresden project's Nuclear Power Group, the Parr Shoals project's Carolinas Virginia Nuclear Power Associates, the Peach Bottom project's High Temperature Reactor Development Asso-

ciates, and the Pathfinder project's Central Utilities Atomic Power Associates. An alternative method of organizing utility group projects is to form a new corporation composed of the participating electric companies. An example of this is the Yankee Atomic Electric Company,[20] which was among the first utilities to generate electric power from nuclear energy.

Another factor leading to the formation of these groups in the early stages of nuclear power was the desire of the utilities to learn about the operation of atomic facilities from direct experience. Since there were more interested utilities than nuclear projects, such groups provided the opportunity for many electric utilities to learn about nuclear power generation collectively. Rather than having one utility accumulate nuclear experience by financing and conducting all the R&D and construction work, these tasks were performed by a group of utilities.

The operating experience gathered at one type of nuclear power plant usually can be employed at other such plants. Thus, an electric utility that is considering plans to build a nuclear power plant has a definite incentive to join a utility group that currently is constructing a similar power plant. A possible result of having a number of utilities gain nuclear experience is the greater and faster transfer of nuclear power technology. The group research projects have provided the utilities with the information and experience required for making the decision whether or not to build a nuclear power plant.

The role played by group R&D projects in the total electric power industry research program has been limited mainly to the nuclear field. As a paradigm for privately sponsored nuclear power R&D, the above-mentioned utility groups and others have made a considerable contribution to utility-sponsored R&D. When one remembers that many utility personnel believe that the industry's future lies in the growth of nuclear generating capacity, these multiutility efforts may increase in significance.

Industry Associations

Formal industry associations have been one of the fastest growing segments of the electric utility industry's research activity over the past twenty years. Unlike the cooperative efforts of the utilities, the industry associations have been organized on a continual basis. Such organizations as the Edison Electric Institute (EEI), the Association of Edison Illuminating Companies, and the American Public Power

Association (APPA) are well-established industry groups. Until recently the major concern of these organizations was to increase the economic welfare of the industry by promoting increased kwh sales and overcoming the competition of other energy sources. The industry associations, like the utilities they represented, firmly believed that the best, most advanced, and strongest society uses the greatest amount of electric energy per capita. This goal, continued growth of energy consumption, has begun to be criticized within the industry. The utility executives, such as chairman Charles P. Luce of Consolidated Edison, who have questioned the worth of growth and its implications for the economy and environment, nonetheless remain a minority.

At times these trade associations seem to be working at cross purposes. The private (often represented by the EEI) and public (the APPA) sectors of the electric utility industry have conducted vigorous exchanges about the relative merits of their respective ownership. These open debates appear to be buried now and largely have been replaced by cooperative efforts to defend the industry against the growing environmental criticism. This defense has begun to include sponsorship of common R&D projects.

Of the industry associations, the EEI historically has made the largest R&D contribution, although small in comparison to that of the manufacturers or utilities. The EEI represents 182 privately owned electric utilities which serve 76 percent of the nation's electric customers.[21] Most of these companies are relatively small systems for which the EEI Research Division sponsors and conducts the R&D that they cannot individually afford.

The EEI, on behalf of its membership, commenced its coordinated research activities by establishing its Research Projects Committee in 1952. The Research Division's budget initially was small and grew to $238,000 by 1960.[22] The EEI Research Division was formally established in 1961. The research budget, representing yearly expenditures on supported or sponsored projects, expanded to $757,000 by 1963,[23] which represents an insignificant fraction of the total revenues of the utilities constituting the EEI.

By 1963 the estimated total cost of the EEI-sponsored projects was $8.5 million.[24] Of this figure, the EEI had contributed $1.7 million to its thirty-two active research projects. From 1964 to 1967, $3.9 million was budgeted to be spent on these projects. The total cost of the EEI-sponsored or -supported projects increased quickly. By 1969 the worth of similar research projects was over $43 million.[25] Contributions for these projects have been made on an industry-wide basis, with the manufacturers, member utilities, and universities supporting the Research Division's program.

The EEI Research Division has sponsored projects in many areas of electric power generation, transmission, and distribution including research into tree growth control, extra-high voltage cable, underground transmission, and high-temperature steam generation. Consequent to environmental considerations, the EEI has launched research projects examining the opacity measurement of stack plumes, control of sulfur dioxide (SO_2) and nitrogen oxide (NO_x) emissions, and thermal pollution of streams.

Longer-term research projects involve studies into the development of an economical breeder reactor, investigation into the use of nuclear fusion, and direct conversion of electric energy through magnetohydrodynamics.

These and other projects are financed by taxing the EEI's member companies in proportion to their sales of electricity. For 1972 this charge for research expenditures was .033 mills per kwh, a rate which provided an estimated $7 million. The assessment continuously has risen throughout the EEI Research Division's history; since 1968 the charge has quadrupled. In 1974 much of the EEI research program was transferred to the Electric Power Research Institute (see page 48) and was financed by a .10 mills per kwh charge. Research funding of $119.5 million was sought from the privately owned utilities.[26]

The EEI and, to a lesser extent, the other industry associations perform another important research-oriented service for the electric utilities. Information about the nature of utilities' R&D activities is disseminated through the publication of on-going project titles and participants. Industry sources cite as one of the historic weaknesses of the industry's R&D programs a lack of coordination among the utilities' research efforts.[27] Too often information about specific projects was not available to nonparticipating companies, causing inevitable duplication of some research.

The electric utility industry, realizing the severity of this problem of communication and coordination, set about to solve it by two means: The EEI began publishing a list of projects undertaken by member companies to be made available to all utilities. The Electric Research Council, representing all segments of the industry, was formed in 1965 to initiate and coordinate industry R&D projects.[28]

In 1956 the EEI published *Current R&D in the Electric Utility In-*

dustry. This report, and others updating the original, showed what
research by the privately owned utilities was being conducted. Although many of their projects are limited in scope, these utilities have
conducted a large number of research projects. These EEI publications also illustrated that, at least in the mid-1960s, the privately
owned electric utilities did not always communicate among themselves about their research.[29] The problem has yet to be fully resolved.

The issue of communication existed not only among the privately
owned utilities, but also between privately and publicly owned companies. In September 1963, when it appeared that the utilities were
not correcting this problem by their own actions, Joseph C. Swidler
met with representatives of all industry groups to establish a permanent industry-wide research organization. The Electric Research
Council, formed from this meeting, was comprised of twelve utility
representatives—eight from privately owned utilities, and one each
from the Department of the Interior, TVA, the APPA, and the National Rural Electric Cooperative Association.[30]

The ERC acted to promote research beneficial to all the industry
through its initiatives and coordination of specific R&D programs.
Supplanting the EEI's *Current R&D in the Electric Utility Industry,*
the ERC began to publish its *Survey of Research* to inform the industry of on-going research activity.

In the fall of 1970 the ERC established an R&D task force to define
the utility industry's future R&D goals and estimate their cost. This
task force published a report in September 1971 entitled *Electric Utilities Research and Development Goals Through the Year 2000,* which
outlined hundreds of R&D projects that the task force thought needed
to be undertaken. To finance this ambitious power industry R&D program, the ERC report called for yearly R&D expenditures of $1.12 bil-
lion from government, manufacturers, and privately and publicly
owned electric utilities.

A new industry-wide R&D organization, the Electric Power
Research Institute (EPRI), was incorporated in March 1972 to administer, organize, and implement the electric utilities' portion of the
R&D program outlined in the 1971 task force report. The pledged
R&D funds for 1974 coming from the member utilities totaled $95.5
million, with another $23.4 million possible. This pledged total
represents a 56 percent rise over the 1973 estimated total R&D expenditures. About one-third of the total R&D expenditures have been
channeled to nuclear power research—mainly work on the liquid

metal fast breeder reactor (LMFBR) project and fusion research. EPRI planned to assign $25 million annually, beginning in 1973, to the Commonwealth Edison-TVA-AEC LMFBR demonstration plant.

EPRI has succeeded the EEI Research Division and the ERC as the utility industry organization responsible for providing the direction and financial support for industry R&D projects. Initially, EPRI-sponsored and -funded R&D will be contracted to existing laboratories and research facilities. Later on, EPRI may construct and staff its own research laboratories.

To a large extent EPRI is the end result of intense public and private pressure placed on the electric utility industry to create and administer its own effective, responsible R&D program. Beginning with the EEI's efforts in the 1950s, then moving to the first public and private utility-coordinated R&D efforts with the ERC, the utility industry has now established the EPRI to represent it in the important and increasingly visible area of R&D activity. If this relatively new organization is to be successful, all sectors of the utility industry together with the manufacturers and government will need to interact in a harmonious, coordinated, and continuing effort designed to solve the complex problems of creating an efficient, reliable, and environmentally safe energy system. To date the industry's R&D efforts have been too limited and too ineffectual to hope for any such success. Solutions to the industry's R&D problems still may be difficult to resolve, but with the industry's commitment to confront them, the chances of success now seem more likely.

Colleges and Universities

The role performed by the colleges and universities in electric power R&D and innovation is distinct from that of the manufacturers, utilities, or industry associations. The universities rarely initiate their own R&D projects, instead they perform research that the industry groups or government request and finance.

The universities have conducted "a significant fraction"[31] of industry research, although this amount is far less than that of manufacturers or utilities. It includes projects undertaken on the campus by students and by engineering faculty members. The source of funding is industry-wide, including all of the above-mentioned groups.

Types of R&D performed by colleges and universities range from large, theoretically oriented examinations to small projects requiring detailed scientific analysis. As expected, they conduct R&D that is tailored to their engineering department programs: fundamental, basic research, and laboratory inclusive projects that are beneficial in educating graduate students.

Examples of research projects performed by universities (in this case sponsored by the EEI) are the following: the extension of steam tables to cover higher temperatures and pressures,[32] Brown University and California Institute of Technology; the supercooled generator project, Massachusetts Institute of Technology; high-temperature metallurgy research, University of Michigan with Battelle Memorial Institute; thermal pollution of streams, Lehigh University; electric resuscitation, Johns Hopkins University; extra-high voltage cable, Cornell University; and high-temperature steam generation, jointly conducted at Philip Sporn Station by Purdue University and the U.S. Navy Laboratories.[33]

Such research funding was not always forthcoming. University-conducted engineering R&D was power oriented prior to the early 1950s. But with the birth of large, government-sponsored R&D programs such as NASA, power industry R&D became overshadowed and was given a lower priority. Reasons for this change were found both in the industry and the universities. As NASA R&D funding to universities rapidly increased, the schools began to move away from power-oriented research into the more "glamorous" areas of aerospace R&D. The electric power industry retreated from the academic arena and cut its R&D fund allocation to the universities. Power research on the university campuses dried up for ten years after the mid-1950s. Since then the industry, largely through the EEI Research Division and the federal government, has again financed power research on the campuses.[34]

During this period when industry-sponsored R&D was small in supply, the number of universities teaching power engineering and related fields decreased. This greatly reduced the supply of college-trained personnel both interested in and qualified for employment by the electric utilities. In time, the utilities began to recognize the growing manpower problem. Articles and papers were written extolling the career opportunities present in the industry. The utilities attempted further to convince the engineering students by "selling" themselves through an expanded R&D program to be performed by students at the universities. The electric utilities reasoned that "an important by-product of exposing graduate students and professors to the more glamorous aspects of utility engineering should be a specific part of this [R&D] program, helping to dispell false impressions which are claimed to have turned promising students from utility careers."[35] The decrease in aerospace R&D budgets, taking some of the glamor out of this area of study, assisted in this goal of the utilities.

Since the mid-1960s, the universities have received major research assignments on a project-by-project basis. This type of research usually is suitable to laboratory or mathematical solutions not involving heavy capital expenditures. The future pattern of university-conducted R&D for the electric utilities is not expected to change greatly.[36]

The Utilities' Motives to Perform Technological Activity

The innovative activity most often undertaken by the electric utilities involves the first commercial installation of electrical equipment designed, developed, and produced by the electrical equipment manufacturers. The risks associated with such innovations in operation arise from using equipment untested under actual operating conditions and thus more susceptible to unscheduled outages. The utility's decision to innovate accounts for such risks and weighs them against the potential rewards accruing from the use of innovative equipment. The utility's several motives to innovate, described below, may be personal or economic in nature.

(1) Many of the industry-sponsored R&D projects that have led to commercial innovation are promoted to enable the electric utilities to compete successfully with other energy producers, the gas utilities in particular. (2) Some utility executives engage in R&D to create new uses for electricity, increasing the demand for their product. (3) Anticipated reduction of operating costs and increases in service reliability motivate many utilities to participate in R&D and to innovate. (4) Utility managements may assign personal and company prestige to technological performance created by innovative activity. These firms may be motivated to conduct R&D and innovation more as an end in itself than as a means of achieving other goals.

Interfuel Competition

The electric utilities can and often do compete with other energy producers for industrial, commerical, and residential sales. Since its beginning, the electric utility industry has labeled the gas industry as its principal competitor. Edison's purpose in planning and designing the first electric system was to produce an alternative form of energy for some lighting that would compete with the gas industry.[37]

Since 1892 the market for gas lighting has deteriorated. Currently, interfuel competition is strongest in the markets for industrial sales and residential and commercial space conditioning (heating and cooling). The utilities have launched large advertising compaigns to promote the sale of electricity for these uses. In 1970 the electric utilities reported spending over $67 million on advertising used in part to promote greater sales of electricity—almost 1.5 times the amount spent on R&D.[38]

A growing number of electric utility spokesmen regard research as another means of improving their competitive performance. Speaking at the EEI Convention in June 1961, J. K. Horton declared, "One thing is certain regarding research: We, the electric industry, must do it *first* and *best*—and we cannot afford to let our competitors, the producers of alternative sources of energy, take the initiative from us."[39] Over the past decade the utilities, under pressure not only from other energy producers but also from the public and federal government, have begun to realize that R&D expenditures can lead to direct future benefits for the industry. Lest the reader believe that the electric utility industry is expanding its R&D expenditures by cutting other types of spending, the utilities' sales expenses in 1970 amounted to 1.5 percent of total privately owned electric utility revenues, while R&D expenditures were only 0.2 percent of revenues from 1968 through 1970.[40]

The motive originating from industry interfuel competition best explains the rapid growth of industry association-sponsored R&D. It is strongest in organizations with industry-wide membership. Acting for its constituents, the EEI Research Division is interested not only in research for its own sake, but also as a way of providing market security for the electric companies in the face of interfuel competition. There also may be a desire on the part of privately owned utilities to "outdo" the publicly owned sector to demonstrate their superiority.[41]

Increase in Demand for Electric Power

Since regulation requires electric utilities publicly to justify rate changes before adjusting the selling price of their product, total revenue is more dependent in the short run on changes in quantity sold than on price. Because of this restriction the utilities have long emphasized the need for continual sales growth to ensure adequate industry profits. This can be accomplished by enlarging their service markets and increasing the customer's utilization of electricity.

Some utility officials see a goal of R&D expenditures as increasing the utilization of electricity. For example, "Tennessee Valley Authority Manager G. O. Wessenauer . . . has called for 'some energetic research to find new applications for electric energy.' Noting that no major electric appliance has been invented in the past quarter century, Wessenauer at a recent American Public Power Association workshop, urged public power men to find new ways of cooperating in R&D to develop new utilization devices."[42] Many utility executives in the private and public power sectors, according to interviews conducted by *Electrical World,* felt that more utility R&D was needed because such a load-building (electric power output raising) appliance had not been created. The motive for spending money on research projects, according to these persons, is to increase the growth of the electric utilities' electric power output and revenues.

Even without a new "major electric appliance" being invented, the utilities' growth has far from stagnated. The kwh sales of utilities have grown at an annual rate of 8.3 percent from 1950 to 1970. The utilities' growth does not seem overly dependent upon the creation of such an appliance.

Not only have the utilities grown rapidly over the past two decades, but also within the past decade power shortages and "brownouts" (voltage reductions to stretch the available power supply) have become more common as utilities have been unable to meet the increasing peak demand for electric power. These shortages have thus far been most prevalent in metropolitan areas, such as New York, Chicago, and several cities in Florida. Among utilities, Consolidated Edison of New York City was first in asking its customers voluntarily to limit their usage of electric power with its "Save A Watt" campaigns. These advertising campaigns were unique in the utility industry in that funds were spent with the intention of limiting the demand for electricity rather than trying to raise the demand.

Operating Costs and Service Reliability

The utilities' desire to reduce operating costs is another motive to undertake R&D projects and innovative activity. According to industry personnel, innovations often occur from the utilities' own insistence upon increasing operating efficiencies by lowering fuel cycle costs, a major cost in the generation of electric power. Sometimes minor changes in the plant operation or design can save large sums of money over the plant's lifetime (plant and equipment usually are amortized over twenty to thirty years). These improvements arising from R&D become economically significant to the utilities.

William Einwechter of the Philadelphia Electric Company stated that his company's primary motive for pioneering the commercial use of 5,000 psi operating conditions was to increase thermal efficiency (to over 40 percent), and thus lower fuel cycle costs. Larry Dwon of AEP, echoing the opinion of other utility managements as well, stated that "the drive to supply energy at the least possible cost is the major motive to innovate."[43]

Utilities also may innovate by operating pioneering equipment, not only to lower current operating costs, but also to stimulate more research into methods of decreasing future costs. For example, Carolina Power & Light designed its Goldsboro Plant Unit 3 for automated operation to improve plant efficiency and reliability, and to restrain the rising costs of plant construction, operation, and maintenance. "Conservative estimates of the worth of these benefits did not quite justify the cost of automation. . . . But it appeared that automation at present would lay the groundwork for substantial savings in plant costs, particularly when it has advanced to where conventional controls can be omitted."[44]

One of the major economic concerns of the utilities, along with the price of electricity, is the reliability of their service. The utilities correctly believe that the public measures a utility's record as a business through the dependability of the electricity supply. Because customers complain vociferously when electricity is unavailable,[45] the utilities have paid close attention to the reliability of service. Even though electric power is available 99.8 percent of the time,[46] the electric utilities appear fully committed to lowering that 0.2 percent.

Some utilities have in the past decided to purchase "less innovative" equipment because the conventional, proven machinery is more reliable. As may be expected, the newer the design of the equipment, the more likely it can cause operating problems. It is no surprise to read the opinion that "the primary purpose of research carried out by the electric power companies is to improve the quality of service and to maintain the price level of electric energy in the face of rising fuel and labor costs."[47]

Corporate and Personal Prestige

Although hardly an economic motive, enhancement of corporate and personal prestige is at times a powerful force persuading both manufacturers and utilities to pursue R&D ventures. If one subscribes to the belief that individuals, not organizations, are responsible for most pioneering decisions and innovations, then this motive is a dominant one in the R&D process. The following are examples of its workings and results in the electric power industry.

The contributions and impact of Samuel Insull on the electric power industry have been numerous. It was he who did much to change the managerial and financial structure of the electric utilities in the 1920s. Another technical plunger was Samuel Ferguson of the Hartford (Connecticut) Electric Light Company. Like Insull, he created an atmosphere of experimentation at his company and at other firms. Over the past two decades the man who is most often mentioned as the spokesman for the progressive, technically oriented segment of the utility industry is Philip Sporn. Through his efforts and those of other similarly oriented persons, he brought the American Electric Power Company to a position of industry leadership. The fulfillment of ideas fostered under his leadership created pioneering developments such as the first supercritical pressure steam unit and the high-voltage transmission grid. He foresaw the need for interconnections of load centers with high-voltage transmission lines, and the use of larger units with promotional electric rates, to draw industrial customers. These additional sales created more savings as scale economies were realized.

Sporn's and the AEP's record has impressed upon the industry that pioneering use of power engineering can solve many of the technical and economic problems facing the electric utilities. He and others believe that the fruits of technological adventureship far outweigh the returns of technological followership. Not all persons in the electric utility industry, or for that matter in any industry, share this feeling. But the technological performance of AEP cannot be disputed, as it has led in many of the developments of power generation and transmission before and after 1950. AEP has established itself as one of the most technologically and economically successful companies in the utility industry.

Walker L. Cisler, president of Detroit Edison, is another technologically oriented utility executive. Detroit Edison has made technical contributions throughout its history, the most prominent being in the field of nuclear power. Cisler has been one of the driving forces behind such experiments. Detroit Edison and its Atomic Power Development Associates' Enrico Fermi sodium-cooled fast breeder reactor was a pioneering industry achievement. What makes the Fermi experiment important, despite its failure to achieve all its assigned goals, is that Detroit Edison and the utilities comprising APDA, with research assistance from the AEC, had to further develop and refine the amount of known information about fast breeder reactors and then employ this new technology to design, build, and operate the Fermi reactor.

The publicly disclosed explanations for Detroit Edison's decision to first build a privately owned fast breeder were to obtain power economically from nuclear energy and to free the United States from reliance upon uranium imports.[48] In addition, there was the challenge and allure of accomplishing something that had not been done before on that scale, and which would benefit the industry and the nation. Detroit Edison, under Cisler's guidance, accepted this challenge and began to sell the idea to other utilities, many of which initially were unsure if it could be done. By making this commitment, both Cisler and Detroit Edison have gained standing and recognition as industry leaders.

In addition to AEP and Detroit Edison, other electric utilities cited as being research oriented are Commonwealth Edison (one of the companies once in Insull's empire), Consolidated Edison, Cleveland Electric Illuminating, Pacific Gas and Electric, Philadelphia Electric, and Southern California Edison. In many of these companies the influence of one executive has proved the decisive factor in choosing to adopt a technically oriented business strategy. These individuals have used techonological performance as a measure of their company's integrity.

What characteristics are common to such men? The self-made individualist played an important role in the history of many electric utilities. These individualists, like Insull, valued competition more highly than cooperation with one another. This competition took technical as well as economic forms. Although the utility executives who acted in this manner were, and probably still are, few in number, they have exerted a disproportionate influence on their industry. According to K.W. Miller of the Armor Research Foundation, there was and still may be technological competition between these research-oriented utilities.[49]

In the opinion of knowledgeable observers, such a rivalry for technological advance existed between AEP and Philadelphia Electric, whose presidents, Philip Sporn and R.G. Rincliffe, both were intent on advancing plant thermal efficiencies by increasing operating pressures and temperatures. Under Sporn, AEP pioneered the use of supercritical temperatures and pressures. Philadelphia Electric later built and operated a unit (Eddystone Unit 1) at 5,000 psi and 1,200° F, with a 42 percent thermal efficiency, the highest commercial operating conditions and efficiency to date. In addition to enhancing company prestige, such rivalry advanced the progress of the industry as a whole.

To these men the process of technological innovation is a fully competitive one, where various firms each are seeking to better their performance and reputation through technological achievement. The competition revolves around which firm is the first successfully to accomplish and perform a technological advance. Such competition is by no means widespread in the electric utility industry. When asked what management factors have greatly affected the process of technological innovation in the industry, Philip Sporn replied, "Perhaps the most important of these is the lack of a competitive spirit, a feeling on the part of those in the industry that it does not really matter whether technological innovations and advances are developed."[50]

Individuals like Sporn feel that the risks involved in such undertakings as R&D and innovation are favorably weighted by the rewards accruing to themselves, their companies, and the industry. Although they regard the quality of their company's economic record as necessary and important, they place emphasis on the record of technological achievement as another measure of the firm's efficiency and progressiveness. Men such as Sporn, Rincliffe, Cisler, and others may view this motive involving company prestige as the strongest one of all.

The next chapter completes the descriptive analysis of the electric utilities' innovative activity in electric power generation. It presents the record of power generation innovations and then describes those innovations judged to be the most technologically significant.

IV

Technological Innovations in
Electric Power Generation
1950–1970

The preceding chapters examined the process of technological innovation in the electric power industry, its participants, their motives, and the factors affecting this process; the present chapter documents and describes the products of the process, successful technological innovations. Appendix B summarizes the process of producing electric power for the reader unfamiliar with the technical aspects of steam-electric power generation.

Sixty technological advances initiated by various electric utilities are listed in Table 4.1 as evidence of technological innovation in electric power generation between 1950 and 1970.[1] Only those innovations publicly disclosed in the sources listed in chapter I are included. Although the list may not be all inclusive, industry personnel who critically examined it for accuracy and completeness did not indicate that any significant advances were overlooked. Following Table 4.1 is a description of the most important innovations classified by the type of innovative activity in power generation displayed by the electric utilities.

Table 4.1. *Selected Technological Innovations in Electric Power Generation, 1950–1970*

Boilers, boiler design
  [1] First boiler to operate at supercritical pressures (4,500 psi), Philo Station Unit 6, Ohio Power Co., 1957.
  [2] First once-through Sulzer monotube boiler to operate at supercritical pressures in United States, Eddystone Unit 1, Philadelphia Electric Co., 1958.

58

Table 4.1.—*Continued*

[3] First once-through Sulzer monotube boiler to operate at subcritical pressures, Frank M. Tait Station Unit 4, Dayton Power and Light Co., 1957.

[4] First pressurized boiler, Knox Lee Power Plant, Southwestern Electric Power Co., 1950.

[5] First welded-wall boiler, Kearny Station, Public Service Electric and Gas Co., 1952.

[6] First combined-circulation boiler, Bull Run Station, TVA, 1962.

[7] First double reheat steam generating unit, Philo Station Unit 6, Ohio Power Co., 1957.

Boiler feed pumps, turbines, generators

[8] First high-speed boiler feed pump, Kearny Station, Public Service Electric and Gas Co., 1952.

[9] First large installation of main turbine boiler feed pump drive, Astoria Station, Consolidated Edison Co., 1956.

[10] First single boiler feed pump per boiler installation, driven by auxiliary steam turbine, Glen Lyn Station, Appalachian Power Co., 1956.

[11] First brushless exciter, Springdale Plant, West Penn Power Co., 1960.

[12] First installation of conductor-cooling with hydrogen gas of generator rotor, Edgewater Station Unit 3, Wisconsin Power and Light Co., 1951.

[13] First generator with inner-cooling on both rotor and stator, C.R. Huntley Station, Niagara Mohawk Power Corp., 1954.

[14] First large liquid-cooled generator, Eastlake Station, Cleveland Electric Illuminating Co., 1956.

[15] First water-cooled generator in United States, Philip Sporn Station, jointly owned by Ohio Power Co. and Appalachian Power Co., 1959.

[16] First double-rotation turbine in United States, Beacon Street Heating Plant, Detroit Edison Co., 1960.

Condensors, cooling towers, water purification

[17] First large condensor using aluminum tubing, Oak Creek Station Unit 3, Wisconsin Electric Power Co., 1955.

[18] First condensor completely tubed with stainless steel, Rivesville Station, Monongahela Power Co., 1959.

[19] First large unit side-entry condensor (connected to first axial-flow exhaust turbine), Portland Power Station, Metropolitan Edison Co., 1958.

[20] First hyperbolic, natural-draft cooling tower in United States, Big Sandy Unit 1, Kentucky Power Co., 1962.

[21] First canned motor pump for boiler feed water circulation in a closed loop system, Possum Point Plant, Virginia Electric Power Co., 1955.

Pollution abatement equipment

[22] First use of a bag filterhouse to reduce stack emissions from oil-fired stations, Alamitos Station, Southern California Edison Co., 1964.

[23] First use of two-stage combustion system to reduce NO_x emissions, Alamitos Station, Southern California Edison Co., 1964.

[24] Pilot plant use of Wellman-Lord process of SO_2 removal, Cannon Power Plant, Tampa Electric Co., 1968.

[25] First operational tests of Combustion Engineering process of SO_2 removal, St. Clair Power Station, Detroit Edison Co., 1966.

Table 4.1.—*Continued*

[26] First pilot plant tests of Monsanto process of SO_2 removal, Seward Plant, Pennsylvania Electric Co., 1962.

[27] First use of tall stacks for effluent dispersion, Kyger Creek Station, Ohio Valley Electric Corp., 1955.

Automatic control equipment

[28] First central station use of automatic data logging, Neches Power Station, Gulf States Utilities Co., 1957.

[29] First U.S. plant with data logging, scanning, and alarming function control system, T.H. Allen Plant, City of Memphis, Tennessee, 1958.[*]

[30] First U.S. power plant with data logging, scanning, and alarming functions, plus performance calculations, Sterlington Station, Louisiana Power & Light Co., 1959.

[31] First use of transistorized, general purpose digital computer in a U.S. power plant, Sterlington Station, Louisiana Power & Light Co., 1959.

[32] First completely automated steam-electric generating plant, Little Gypsy Station, Louisiana Power & Light Co., 1961.

[33] Pioneering automation of a 2,400 psi, single reheat, once-through boiler steam generating unit, Huntington Beach Plant, Southern California Edison Co., 1958.

[34] First TV viewing of furnace combustion, Port Jefferson Plant, Long Island Lighting Co., 1951.

Gas turbines, combined cycle units

[35] First regenerative-type gas turbine in central station use in the United States, Bangor Hydroelectric Co., 1951.

[36] First fully automatic "self-starter" for gas turbine in central station use, Farmingdale Station, Central Maine Public Power Co., 1950.

[37] First combined cycle, gas turbine-steam turbine unit, Rio Pecos Plant, West Texas Utilities Co., 1954.

[38] First combined cycle use of gas turbine exhaust gases to preheat air for conventional boiler, Warrick Plant, Crisp County (Georgia) Power Commission, 1959.

[39] First coal gas-fired, gas turbine to operate in combined cycle with steam unit, Muskingum River Plant, Ohio Power Co., 1957.

[40] First jet gas turbine repowering of generation units, Waterside Station, Louisville Gas and Electric Co., 1967.

Fuel handling

[41] First commercial conveying of coal by slurry pipeline, Cleveland Electric Illuminating Co., 1957.

[42] First implementation of "integral train" concept of shipping coal, Pennsylvania Power & Light Co. (with Pennsylvania Railroad), to Brunner Island, Pa., plant, 1962.

Operating temperatures and pressures

[43] First 1,100°F steam generation, Kearny Station, Public Service Electric and Gas Co., 1952.

[44] First 1,150°F steam generation, Philo Plant Unit 6, Ohio Power Co., 1957.

[45] First 1,200°F steam generation, Eddystone Unit 1, Philadelphia Electric Co., 1958.

Table 4.1.—*Continued*

[46] First 3,500 psig boiler, Avon Lake Plant, Cleveland Electric Illuminating Co., 1958.

[47] First 5,000 psi boiler, Eddystone Unit 1, Philadelphia Electric Co., 1958.

Nuclear plant and equipment

[48] First commercial nuclear power station, Shippingport Atomic Power Station, Duquesne Light Co., 1957.

[49] First large-scale nuclear power plant, Indian Point Station, Consolidated Edison Co., 1962.†

[50] First private, prototype fast breeder reactor, Enrico Fermi Reactor, Detroit Edison Co. with Atomic Power Development Associates, 1960.

[51] First full-scale nuclear power plant with integral nuclear superheat, Pathfinder Plant, Northern States Power Co. with Central Utilities Atomic Power Associates, 1964.

[52] First organic-cooled and -moderated reactor, Piqua Plant, City of Piqua, Ohio, 1962.

[53] First sodium graphite-moderated thermal reactor, Hallam Power Facility, Consumer's Public Power District, 1962.

[54] First heavy water-cooled and -moderated pressure tube reactor, Parr Shoals Nuclear Plant, Carolinas Virginia Nuclear Power Associates, 1963.

[55] First high-temperature gas-cooled reactor, Peach Bottom I, Philadelphia Electric Co. with High Temperature Reactor Development Associates, 1963.

[56] First jet-type water pumps within the pressure vessel, Dresden II, Commonwealth Edison Co., 1969.

[57] First use in a large pressurized water reactor of a noncanned motor pump unit, San Onofre Plant, Southern California Edison Co., 1967.

[58] First commercial reactor using prestressed concrete pressure vessel, Fort Saint Vrain Reactor, Public Service Co. of Colorado, 1972.

[59] First reinforced concrete pressure containment structure, Parr Shoals Nuclear Plant, Carolinas Virginia Nuclear Power Associates, 1962.

[60] First nuclear power plant to use ice condensor system for nuclear reactor containment, Donald C. Cook Plant, Indiana & Michigan Electric Co., 1973.

NOTES: See Appendix J for the name of the electric utility system affiliated with the operating utilities listed in this table. The date given with each innovation represents the year when the plant or unit began commercial operation, not when the decision to innovate was made.

° Plant is leased by the City of Memphis to TVA.

† Although Commonwealth Edison's Dresden I nuclear power plant was in operation in 1960, Consolidated Edison filed its license application with the AEC before Commonwealth Edison did.

All of the technological innovations in electric power generation accounted for in Table. 4.1 are capital-embodied. Because there is little inducement to replace economically operating equipment, the incentive to undertake innovative activity in power generation is greatest when the utility is either expanding its generating capacity by building new generating units, or replacing worn-out equipment in older generating stations. The majority of innovations listed in Table 4.1 were installed in new electric generating plants.

Although fundamentally new methods of producing electric power have not been commercially employed by the electric utilities within the surveyed period, there have been great improvements in existing generation technology. All of these technological advances either: (1) increased the generating unit's operating efficiency or lowered its operating costs; (2) automated the controls of power generation, thereby providing more reliable thermal performance; (3) reduced the generating unit's environmental impact; or (4) introduced the use of a nuclear-fueled heat source to produce steam. The most technologically significant innovations[2] are described below; the remaining innovations listed in Table 4.1 are briefly described in Appendix C.

Innovations That Increase Thermal Efficiency

The first supercritical unit [1].[3] Under Philip Sporn's leadership, the American Electric Power Company's operating division, Ohio Power Company, pioneered the commercial use of supercritical steam to generate electric power. The electric utility industry generally has agreed that the shift of supercritical pressures and temperatures (where water turns directly into steam without any intermediate stage of "bubbling") is a new power generation milestone.

Ohio Power's Philo Plant supercritical unit significantly advanced steam generation techniques for the following reasons. (1) It was designed to operate more efficiently than existing units. Steam was generated at 4,500 psi and 1,150°F resulting in a designed heat rate of 8,500 BTU per kwh, about 500 BTU better (less) than the then best subcritical units. This higher thermal efficiency reduced fuel cycle costs. (2) It was the first commercial supercritical unit to be designed, constructed, and operated. (3) It provided AEP with the technical and operating experience enabling the company to later design, construct, and operate supercritical units of greater capacity. (4) It was an important step leading to the commercial use of new generating methods by which the electric utilities could meet the public's increased demand for electric power without substantial cost increases.[4]

In their 1953 decision to build a unit using supercritical pressures and temperatures, General Electric Company (builder of the turbine-generator), Babcock & Wilcox Company (builder of the boiler), and AEP sought to reduce operating costs by realizing greater thermal efficiencies. Utilization of the "once-through" boiler design at supercritical pressures eliminates the need for the heavy and costly boiler drums of conventional evaporative boilers. Thus there are savings in operating costs from reducing the net heat rate and from simplifying the boiler design. Also, the capacity of a given-sized unit is increased as operating temperatures and pressures are raised. Partially offsetting the thermal gains derived from supercritical operation is the increased work required by the boiler feed pumps at high pressures.

In steam-electric power generation a close relationship exists between plant design and fuel cycle cost. Increases in the cost of fuel usually prompt electric utilities to seek methods of reducing fuel usage to the economic minimum. This incentive promotes the development of generation methods with higher thermal efficiencies at low incremental investment costs. Supercritical operation becomes economic when fuel costs are relatively high. When fuel is relatively inexpensive, savings from supercritical operation usually are not great enough economically to justify building supercritical units.

The entire Philo supercritical project, which cost over $12 million, was engineered by the AEP Service Company. Unlike most other innovating electric utilities, AEP played an active role in the entire development of the Philo unit and cooperated with the manufacturers in the design of the components. The project was said to be the consummation of more than thirty years of continuous work and development on the AEP system.[5]

The first once-through Sulzer monotube boiler to operate at supercritical pressures [2]. The Philadelphia Electric Company expanded the technological frontier of steam-electric power generation with its decision in late 1953–early 1954 to build the Eddystone Station Unit 1, incorporating a number of innovations. The nation's first supercritical monotube boiler for steam generation attained the highest commercial operating conditions of 5,000 psi, 1,200°F, and double reheat at 1,-050°F. The 325 Mw Eddystone unit was the second commercial, first large-scale supercritical unit (the capacity of Philo Unit 6 was 125 Mw) in the United States to go into operation and one of the most efficient units in the nation: its designed heat rate, 8,016 BTU per net kwh, was very low, and its thermal efficiency was rated at over 40 percent. In 1961, 1962, and 1963 Eddystone was the nation's most efficient fossil-fueled generating plant in terms of average annual heat rate.

As discussed above, supercritical boilers employ a once-through design that simplifies the design of boiler components. The Sulzer monotube once-through boiler design used for the first time in Eddystone Unit 1 does not use a steam drum or steam separator. The feedwater, maintained above the critical temperature, passes through the continuous economizer and the evaporative and superheater circuits. Unlike the conventional subcritical boilers, there is no recirculation of water through the steam generating circuits: it passes through the fluid circuit only once, hence "once-through." This boiler was designed for a primary steam flow of 1.54 million pounds per hour.[6]

The technological competition mentioned in chapter III between Philip Sporn and R.G. Rincliffe, president of Philadelphia Electric, yielded valuable information on types of station design and operating conditions that were feasible and those that were not. Eddystone Unit 1 demonstrated that relatively efficient base load plant operation technically is possible at very high temperatures and pressures. Philadelphia Electric learned, however, that the relative gains of increased thermal efficiency were offset by higher capital cost requirements: the additional expense of operating at 5,000 psi, 1,200°F could not be economically justified. (Eddystone Unit 2 was designed to operate at 3,750 psi, 1,050°F, and double reheat at 1,050°F.)

Much of the increased capital cost was due to the very expensive piping required for the system. The piping, using high-temperature, high-strength alloys such as austenitic steels, cost up to $1,000 per foot. In general, metallurgy remains a barrier to further gains in power station thermal efficiency. The high cost of currently available high-strength materials limits further significant thermodynamic improvement. Until steam temperatures and pressures can be turned upward again, only modest thermal gains from larger units and from refinements in the steam cycle will be possible.[7]

The technological achievement of Eddystone Unit 1 is not diminished by the unit's excessive capital cost relative to its thermal gains. Philadelphia Electric attained its objective of increasing operating efficiency through higher steam conditions. The company and the electric utility industry learned that, until further developments occur, such operating temperatures and pressures will not be economically justifiable. Industry personnel realize that there is equal worth in learning what is and is not possible.

The first subcritical once-through monotube boiler [3]. The Frank M. Tait Station Unit 4 of Dayton Power and Light generated the first subcritical steam in the United States by means of a Sulzer once-through monotube boiler in 1957.[8] The boiler was designed to produce 940,000 pounds of steam per hour at 2,565 psi, 1,060°F. The Sulzer monotube boiler substitutes a small separating chamber for the steam drum of a conventional subcritical boiler. The monotube boiler eliminates the recirculation of fluid through the steam generating circuits, removing the need for circulating pumps and downtakes. As in the Eddystone boiler, the walls of the tubing do not need to be as thick as in natural circulation boilers, allowing more freedom in boiler design and operation.

The first installation of conductor-cooling with hydrogen gas to a turbine-generator rotor [12]. In 1951 Wisconsin Power and Light Company's Edgewater Station Unit 3 used the first electric power generator in which the conductors of the generator's rotor (the spinning part of the machine; the rotating electromagnet) were cooled with hydrogen gas. The major reasons for pioneering better generator cooling systems were to increase the unit's thermal efficiency by reducing the heat loss and to increase the rating (capacity) of given-sized generators, so that more energy could be produced at lower unit capital cost.

The electric utilities and equipment manufacturers recognized the need for more efficient heat removal systems as the generator's rating and physical size increased. Over the past thirty years generator ratings have increased by more than twelvefold, while the size of the generators has increased by less than threefold. The level of net heat to be removed from a given-sized generator has thus increased fourfold during the last three decades.[9] In the 1930s air cooling was universally used. Hydrogen cooling, developed by General Electric in the 1930s, was first employed in commercial production of electric power at Dayton Power and Light's Frank M. Tait Station in 1937 (see Appendix D). According to John Calhoun, manager of Dayton Power and Light's Electric Production Divison, "the main consideration which influenced DP&L management to gamble on this untried generator was the economic factor—lower cost, and what appeared, at the time, to be probably greater reliability."[10] In the first hydrogen-cooled generators, only the stator (the stationary part of the machine composed of a hollow tube lined with wire windings surrounding the rotor) was cooled by the gas. The technological innovation pioneered by Wisconsin Power & Light was to cool the rotor windings with hydrogen. By cooling these windings more effectively, their resistance is reduced, allowing more current to flow through them. More current produces a stronger magnetic field; the stronger the field, the more energy the machine can generate. The generator's efficiency is raised, increasing its rating. Thus a better cooling system increases a given-sized generator's capacity, lowering unit capital costs of power generation.[11]

The first combined cycle, gas turbine-steam turbine unit [37]. The shortcomings of the conventional operating cycle of steam production have been recognized by engineers for a long time. Stated simply, much of the energy of the generated steam is unavailable due to the nature of the vapor. As thermal energy is lost, usually through unrecoverable heat, the thermal efficiency of the steam cycle is lowered. To reduce this loss, combined cycles (utilizing more than one heat-carrying medium) were developed. A pioneering step in using combined cycles was the first mercury-steam cycle unit at the Hartford Electric Light Company in 1922 (listed in Appendix D). This type of combined cycle never gained wide acceptance due to the drawbacks of mercury as a working fluid. As gas turbine technology developed to the extent that its reliability and economy permitted utilities to use gas turbines for power generation, the gas turbine-steam turbine combined cycle was developed.

In a typical gas-steam cycle, both the gas turbine and steam turbine produce electricity. The gas turbine and steam cycle are interconnected, as the heat of the gas turbine's exhaust gases is used to preheat either the steam cycle feedwater or boiler gases. A portion of the normally unrecoverable exhaust heat of the gas turbine is recovered and used to increase the thermal efficiency of the steam cycle. The steam cycle efficiency can be increased by over 15 percent with the use of the gas turbine exhaust for feedwater heating.[12] The first such gas-steam combined cycle plant at the West Texas Utilities Company's Rio Pecos Plant realized this goal of increased thermal efficiency.

The first use of gas turbine exhaust gases to preheat boiler gases [38]. A slightly different gas-steam cycle configuration was pioneered at the Crisp County (Georgia) Power Commission's Warrick Plant in 1959. Instead of heating the steam cycle's boiler feedwater, the gas turbine exhaust was used to preheat the air entering the conventional boiler. Like the Rio Pecos Plant, the Warrick Plant achieved greater thermal efficiency since less thermal energy was lost.

The additional fuel requirements of the gas-steam cycle have prevented its wide applications. Also, impending FPC restrictions on the use of natural gas as a boiler fuel may inhibit further gas-steam cycle usage.

Innovations That Automate Controls

The first U.S. power plant with automated data logging, scanning, and alarming functions, and performance calculations [30]; the first use of a transistorized, general purpose digital computer in a U.S. power plant [31]; the first fully automated steam-electric generating station [32]. Large generating units operating at high temperatures and pressures require the sophistication and reliability of automated, electronic control systems. Two motives have prompted the electric utilities to improve and to automate power plant control systems. (1) Automated controls optimize operating efficiency, safety, and plant reliability. (2) Electronic control systems require fewer operating employees, reducing the unit labor costs of electric power generation.

Completely automated operation of a steam-electric power plant has developed gradually, each step bringing more control functions under programmed, computerized operation. With few exceptions these advances have been developed by the manufacturers of control systems equipment and then sold to the utilities.

One of the first steps utilized a centralized combustion control system in 1922 (listed in Appendix D). Louisiana Power & Light Company, a leading utility in the area of automated control, pioneered the first use of a transistorized, general purpose digital computer in a U.S. power plant at its Sterlington Station in 1959. This computer performed data logging, scanning, alarming, and performance calculation functions. The computer for the first time gave plant operators accurate, current data on the plant's heat rate, operating temperatures, pressures, and fluid flows—information necessary to maximize the unit's thermal efficiency.

Direct computer control over plant operations was pioneered at Louisiana Power & Light's gas-fired Little Gypsy Station in 1961. The control loop was then complete, as the computer could exercise direct control on current plant operations without human intervention. Fully automated control increased Little Gypsy's plant cost by about 2.5 percent over nonautomated, centralized control stations.[13] To Louisiana Power & Light, this extra cost was justified by reductions in labor costs and by increases in operating safety, reliability, and efficiency.

The dollar benefits derived from the use of automated controls have been estimated by electric utility personnel as follows. The fuel savings achieved from accurate, consistent performance calculations vary according to unit size; representative savings are $50,000 to $150,000 per year on units of 250 to 750 Mw (the Little Gypsy Station unit had an initial capacity of 223 Mw). The resultant improved plant reliability decreases maintenance cost, reduces forced outages, and prevents catastrophic failure. Savings of $25,000 to $75,000 per year in maintenance and $20,000 to $60,000 per year for forced outages can be attained; manpower savings of $25,000 to $75,000 per year have been realized from the use of automated controls; control panel savings of $40,000 to $60,000 per year also are possible.[14]

An advantage of automated control systems is that they increase plant load (output) flexibility, allowing the utility to vary the load of the plant to a greater degree than if nonautomated controls are used. Such load flexibility is possible since the controls simplify the plant operators' responsibilities for altering plant output and increase plant operating safety and reliability. Electric utilities generally try to avoid unnecessary variations in a base load fossil-fueled plant's output because the process of changing the output of these more efficient plants is complicated and lengthy. Instead those utilities that have hydroelectric plants on their systems usually rely on these plants' output to meet daily fluctuations in power demand, while keeping the base load fossil-fueled plants' output relatively stable. A hydroelectric plant's load is more easily and more quickly changed than a fossil-fueled plant primarily because the hydroelectric power cycle is far less complicated than the fossil-fueled power cycle.[15]

Utilities, such as Louisiana Power & Light, that have no hydroelectric plants to rely on for their base load variations must vary the output of fossil-fueled plants when daily changes in demand occur. Such utilities could justify the additional cost of automated control systems since the benefits were great. Other utilities that did not have to fluctuate their base load fossil-fueled plants' output may not have benefited enough to warrant the cost of automated control systems and thus chose not to install automated control equipment.

The heat value (BTU per unit of volume) of a steam-electric plant's fuel also influences which utilities are able to install automated control systems. The more variable the fuel's heat value, the more sophisticated the plant controls must be in order to compensate for changes in heat value. The heat value of natural gas is the most consistent of the fossil fuels, while oil's heat value is only slightly more variable. As the use of coal as a fuel presents combustion control problems due to the much greater variation in its heat value, most of the fully automated power plants are noncoal-burning.

Until automated control technology makes further advances, those utilities that operate gas- or oil-fired generating plants will find it more feasible to install electronic, fully automated control systems. Thus the type of power generation facilities and the fuel available to the utility exert a great influence on whether the utility would benefit from installation of automated controls.

Pollution Abatement Innovations

A pressing technological, economic, and social issue facing the electric utility industry is the development and use of adequate pollution control equipment. A significant portion of the nation's air and water pollution can be directly related to the generation of electric power, as power plants annually produce approximately one-half of the sulfur oxide emissions in the United States.[16] Although the electric utilities long have been aware of their generating plants' environmental impact, it was not until the last two decades that a comprehensive examination of pollution abatement techniques was undertaken in the electric power industry.

An increasingly arduous and expensive solution adopted by the utilities to solve pollution problems is to switch to cleaner fossil fuels such as natural gas, low-sulfur oil, or low-sulfur coal. The cleanest fossil fuel is natural gas, but its supply is insufficient for the needs of the utility industry, and its future availability as a power plant boiler fuel is uncertain. In 1970, one-quarter of all steam-electric generating plants used natural gas (accounting for one-sixth of the gas consumed in the nation).[17] Low-sulfur oil and coal also are in short supply and plagued by location problems. Commonwealth Edison in Chicago has purchased its low-sulfur coal from distant Wyoming and Montana and its low-sulfur residual fuel oil from Venezuela.

The gasification of coal (the production of burnable fuel gas from solid coal) has been long sought after by the utilities as a technique to provide an abundant, economical, pollution-free fuel; but despite years of research efforts by the U.S. Bureau of Mines and other public and private organizations, a commercial coal gasification process remains uneconomical. One of the consequences of the rise in the prices of fuels that compete with coal (oil and natural gas) will be to make currently available coal gasification technology relatively more economical and therefore more commercially attractive.

The innovations described below typify the power industry's research efforts to reduce pollution created by electric power generation. They include methods designed to diffuse effluents effectively, to reduce SO_2 emissions, and to lower thermal pollution of the generating plant's cooling water supply.

The first use of tall stacks for effluent dispersion [27]. The Ohio Valley Electric Corporation's Kyger Creek Station, designed by the American Electric Power Service Corporation, began operating in 1955. Pending the development of practical technology to remove sulfur from fossil fuels, the use of tall stacks has been one of the better methods of eliminating ground-level concentrations of sulfur oxides and for providing adequate effluent dispersion. The tall stacks do not eliminate the pollution, but more effectively redistribute it into the atmosphere. The height of each of the three stacks at the Kyger Creek Station as determined by the local atmospheric conditions was 538 feet.[18] The AEP system operates plants with stacks up to 1,200 feet tall.

The first operational tests of the Combustion Engineering process of SO_2 removal [25]; *the first pilot plant tests of the Monsanto process of So_2 removal* [26]. Recent attempts to improve electric power plants' pollution control systems have concentrated on the removal of the oxides of sulfur and nitrogen from the plant's exhaust gases by chemical means. These methods of SO_2 and NO_x removal, although still prototypical, should lead to more effective pollution abatement devices than the use of tall stacks. The utility industry has been forced by public pressure and public statute to begin testing such devices on power generation facilities.

The Combustion Engineering process of SO_2 removal, unlike other processes, recovers no usable by-product and can be used in existing installations. Dolomitic limestone is passed through a conventional coal pulverizer, mixed with the coal, and fed into the boiler. After combustion occurs, the effluent gases enter a dry scrubbing system to remove the sulfur oxides. The initial operational tests were performed at Detroit Edison's St. Clair Power Station in 1966. After completion of these tests, the first full-scale system for removal of SO_2 and dust from stack gases was undertaken at the Union Electric Company's Meramec Station Unit 2, a 140 Mw installation.[19]

The major waste product arising from this process is magnesium sulfate (epsom salt), which is recovered in the boiler's flyash. A practical use for this magnesium sulfate has yet to be found by the utilities using this process. Storage of the material is difficult because of the environmental damage brought about when the epsom salts come in contact with moisture. Until these problems are alleviated, the practicability of using the Combustion Engineering process of SO_2 removal at commercial plants will be limited.

Another chemical method used to remove SO_2 from stack gases is the Monsanto "Cat-Ox" (catalytic oxidation) process, first tested in a pilot plant project at Pennsylvania Electric Company's Seward Plant in 1962. This method uses a catalyst to convert the SO_2 contained in the stack gases to SO_3, which is then condensed out of the gas stream as sulfuric acid. Unlike the Combustion Engineering process, the Monsanto process uses a wet scrubbing system requiring high-temperature precipitators to remove the sulfuric acid from the gases. Following the pilot plant project tests, a prototype SO_2 removal system was built at Metropolitan Edison Company's Portland Station.[20]

Although these and other methods have been developed to solve the pollution problems facing the utilities, few currently are economically feasible. The scrubbing systems and chemical processes remain costly to install and maintain. The electric utilities have predicted that, if required to install such equipment in all existing fossil-fueled plants, they will have to increase substantially the price of electricity to recover the initial capital expense. Utilities may be able to raise part of the necessary capital with sales of tax-exempt municipal bonds. In addition, regulatory commissions have allowed these capital additions into the utilities' rate bases where they can earn a just and reasonable return.

The first hyperbolic natural-draft cooling tower at a U.S. power plant [20]. Another important area of a power plant's environmental impact is thermal pollution of the plant's cooling water source, usually a river or lake. A significant development in thermal pollution control was the first U.S. power station installation of a hyperbolic natural-draft cooling tower at Kentucky Power Company's Big Sandy Plant Unit 1 in 1962.

Hyperbolic natural-draft reinforced concrete cooling towers have long been used at European power stations. Towers of this type were installed at the Lister Drive Power Station in Liverpool, England, in 1925.[21] The abundance of water resources near U.S. power plants obviated the need to supplement natural cooling methods until recently when the size of generating plants increased relative to the available amount of cooling water. Federal and state regulations specifying the acceptable water temperature differential between inflowing and outflowing cooling water also motivated the utilities to explore ways of stretching the existing water resources.

Cooling towers cool the water that has been heated during the steam-condensing process in the power generation cycle by increasing the surface area between the water and the cooling air. This can be done either by breaking up the water into droplets or by spreading the water in thin films over a large area. By forming a closed-cycle water circulation system, cooling towers eliminate discharges of heated water into the cooling water source.[22]

In contrast to natural-draft cooling towers, induced-draft cooling systems require fans to force cooling air over the water, and cooling ponds require more physical space. The American Electric Power Company expected that the higher initial capital cost of installing a hyperbolic natural-draft cooling tower at the Big Sandy Plant would be offset by savings in piping, fan horsepower, and space required by these alternative thermal pollution control methods.

Nuclear Power Innovations

In generating power from atomic energy, the nuclear reactor replaces the conventional fossil-fueled boiler as the source of heat. The types of reactors differ according to their coolant, moderator, and fuel. There are two designs of light water-cooled reactors, pressurized water reactors (PWR) and boiling water reactors (BWR). Other reactor types are gas (helium) cooled, liquid metal (sodium) cooled, heavy water cooled, and organic (hydrocarbon) cooled. Moderators, the substance used to slow neutrons from the high speeds at which they are released in fission, may be light water, as in a PWR or a BWR; graphite, as in a gas-cooled reactor; or heavy water, as in a heavy water-cooled reactor. The fuel for reactors may be fully or partially enriched U-235, plutonium, thorium, or a combination of these fuels. The vast majority of U.S. power reactors in current use or under construction are light water-cooled and -moderated reactors.

The commercial utilization of these U.S. nuclear reactor designs has been founded on the AEC's experimental reactor program. The AEC reactor testing facilities at Arco, Idaho, Oak Ridge Laboratory, and Argonne National Laboratory were used to construct experimental reactors of various designs to gather information about their operating characteristics and potential for commerical use.

In order to gain a share of the market for power station equipment, nuclear steam supply systems have had to compete favorably with fossil-fueled power generation facilities in terms of total energy costs. Since the mid-1960s the electric utilities have begun to purchase significant quantities of nuclear capacity for their future system needs. Part of this shift from fossil-fueled plants was prompted by the public's growing environmental concern about air and water pollution. Nuclear plants have not been exempt from the environmentalists' concern due to their potential thermal pollution and nuclear waste storage problems, but nuclear plants do not produce any air pollutants since no combustion takes place.

The utilities also have been attracted to nuclear power because of potential cost advantages arising from production economies of scale. As mentioned in chapter II,[23] the manufacturers and utilities believe that nuclear plants enjoy substantial production economies of scale, allowing utilities economically to plan for 1,000 Mw nuclear units. Such attractions have resulted in a sizeable commitment on the part of the utility industry toward a nuclear-based energy system. In 1974 there were a total of 251 nuclear plants either in operation, under construction, or being planned. These plants have a total capacity of over 250,-000 Mw.[24]

When a utility installs a nuclear power facility rather than a fossil-fueled plant, it is substituting capital for fuel. A nuclear plant's initial capital cost is higher than that of a comparable fossil-fueled plant; but over the nuclear plant's lifetime, unit fuel costs are expected to be lower than fossil fuel costs. The lower nuclear fuel cost partially is due to the reduced cost of transporting nuclear fuel: depending on the plant site and other factors, transportation cost may account for one-half of the final cost of fossil fuel. In addition, unlike fossil fuels, some portions of nuclear fuels can be reprocessed and used again.

It is true also that nuclear fuel costs have remained relatively low because of a variety of direct and indirect government subsidies. These subsidies include: (1) a five-year waiver on lease charges for fuel—the government, until recently, has owned the nuclear fuel used by electric utilities and required the users to pay low interest charges, if any at all, on the committed fuel inventories; (2) low guaranteed charges for reprocessing irradiated fuel; (3) high guaranteed buyback prices for the by-product plutonium manufactured during the fission process; and (4) charges for disposing of the nuclear waste that are borne by the government.

The example given in Table 4.2 illustrates the relative differences in fossil-fueled and nuclear power plants' total energy costs. As men-

tioned above, the nuclear power plant's capital cost is higher than that
of the fossil-fueled plant, but this cost disadvantage is offset by the nuclear plant's lower fuel cost. The example illustrates the conditions under which the total energy cost of this nuclear plant is competitive
with the fossil-fueled plant.

Table 4.2. *Comparative Energy Costs for Fossil-Fueled and Nuclear Electric*
Power Plants

		Coal-fueled		Nuclear
Capacity (Mw)		615		605
Heat rate (BTU/kwh)		9,087		10,473
Unit capital cost ($/kw)		107		139*
Fuel cost (¢/million BTU)	20	25	30	—
Energy cost† (mills/kwh):				
capital	2.07	2.07	2.07	2.68‡
fuel	1.73	2.16	2.60	1.45
operation-maintenance	.30	.30	.30	.35
nuclear liability				
insurance	—	—	—	.20
Total energy cost	4.10	4.53	4.97	4.68

SOURCE: D.W.R. Morgan, Jr., "Trends in Thermal Power," Pacific Northwest
Trade Association, 52d Conference, 14-16 September 1964, table I.
NOTE: Figures for coal-fueled plant based on Cardinal Unit (Ohio Power Co.,
began operation in 1967). Figures for nuclear plant based on Oyster Creek Unit
(Jersey Central Power & Light Co., began operation in 1969).
* Includes $3/kw allowance for a switchyard which is already included in the
$107 figure for coal-fueled cost.
†Based on 80 percent capacity factor, 13.5 percent annual charges.
‡ Levelized by present worth techniques.

In this example, when the fossil fuel cost per million BTU rises over
$.25, the nuclear power plant's total energy cost is fully competitive
with that of the fossil-fueled plant. Generalizing from this limited information that nuclear power is fully competitive with fossil-fueled
power generation, of course, is unwarranted, as local conditions and
plant design usually determine which method of power generation is
more competitive. But Table 4.2 does show that commercial nuclear
power technology has developed greatly since its inception at Shippingport Atomic Power Station in 1957, where the estimated total energy cost was 60.15 mills per kwh.[25]

The first U.S. commercial nuclear power station [48]. Nine groups[26]
bid on the AEC contract for the first commercial nuclear power plant.
The contract was awarded to Duquesne Light Company in 1953. The

AEC built, operated, and owned the reactor; the utility purchased the steam produced in the reactor and generated electric power using a turbine-generator that the utility constructed at the site. The Duquesne Light–AEC Shippingport Atomic Power Station began commercial operation in December 1957. Its estimated total cost was $182.9 million, including $110 million for R&D by the AEC, $.5 million for R&D by others, $18.8 million for construction of the reactor, and $12.1 million for construction of the turbine-generator and other plant equipment.[27]

The Shippingport reactor, a 60 Mw PWR[28] manufactured by Westinghouse, was built to test the feasibility of using nuclear power under commercial operating conditions and to gain more knowledge about nuclear power generation. The total energy cost, excluding R&D, was $1,220 per kw.[29] The size of the plant was too small to realize economic unit energy costs, and the plant's operating efficiency was a very low 26 percent, under two-thirds that of the best fossil-fueled plant. A PWR was chosen because more experience and information had been gathered about PWRs (much of it from the Navy reactor program) than about other reactor designs. Although the Shippingport station was not the world's first atomic power station, it pioneered the commercial generation of electric power from light water reactors. (The British had begun nuclear power generation a year earlier at the Calder Hill Station.)

The first high-temperature gas-cooled reactor [55]. Besides the first commercial PWR at Shippingport and the first operating BRW at Commonwealth Edison's Dresden I Station, interest developed in examining the commercial use of enriched uranium, gas-cooled reactors. A promising feature of gas-cooled reactors is their ability to produce superheated steam at a quality comparable to that of large conventional steam stations. Due to their relatively low operating pressures, these reactors could be built in large sizes. The fuels used, enriched uranium with thorium (the fertile material), are efficiently convertible to fissionable U-233. The high fuel-conversion gave these reactors the prospect of obtaining better fuel utilization than other reactor designs, thereby reducing fuel cycle costs.[30]

The first commercial high-temperature gas-cooled reactor was the Peach Bottom I Reactor, operated by the Philadelphia Electric Company and built with assistance from the High Temperature Reactor Development Associates.[31] The AEC and Congress approved the project in June 1959. The members of High Temperature Reactor Development Associates agreed to contribute $16.5 million toward the R&D costs of the project, of which Philadelphia Electric pledged $1.3 million.

The Peach Bottom I station, as a prototype, was built to provide information for the development of larger, more advanced reactors of similar design. The 40 Mw capacity plant was constructed to use a graphite moderator and a coolant of inert helium gas. In operation, the helium is heated to over 1,300°F as it circulates around the fuel elements in the reactor core. The heat is transferred from the helium to the water cycle by means of a heat exchanger. The generated steam is then maintained at 1,000°F and 1,450 psi. Electric power was first produced from the Peach Bottom reactor in January 1967 after a series of delays caused by various problems necessitating design changes. The reactor's net operating efficiency was 34.7 percent,[32] demonstrating that a full-sized commercial, high-temperature gas-cooled reactor could be expected to operate in the 30 to 45 percent thermal efficiency range.

The first private fast breeder reactor [50]. Throughout the AEC's Civilian Power Reactor Demonstration Program, emphasis has been placed on the development of a fast breeder reactor.[33] In fact, the breeder has remained the keystone to the commercial generation of nuclear power; many industry personnel believe that breeding must be successful before nuclear power can become economically competitive with fossil-fueled power.[34]

In April 1952 the Dow Chemical Company and the Detroit Edison Company, after concluding a one-year study for the AEC examining the feasibility of using atomic power for industrial purposes, submitted a proposal to conduct an R&D effort on a power breeder reactor. In March 1955 Detroit Edison and the newly incorporated Atomic Power Development Associates (APDA)[35] proposed that they design, build, and operate an experimental fast breeder reactor.

The motives that prompted Detroit Edison to work on development of the fast breeder reactor have been examined in chapter III. The objectives of APDA were broadly defined:

(1) To study the potentialities of economic production and utilization of atomic energy for the generation of electric power and for other useful purposes. (2) To engage in research, experiments, and development activities, and to contract for work to be done by others, looking toward the practicable industrial utilization of atomic energy. (3) To acquire and provide information on industrial and other civilian uses of atomic energy.[36]

Out of APDA, the Power Reactor Development Company was formed to construct the reactor. Sodium was chosen as the coolant for its properties of liquidity at low temperatures, a very high boiling point, a reasonably low neutron absorption cross-section, and excellent heat-transfer and heat-transport characteristics. These qualities enhance the possibility of realizing high operating temperatures and high thermal efficiencies. The fuels used were plutonium and U-233. Fast breeder reactors do not use any moderator, as high neutron speed is desired. The problems faced by such a reactor design included rapid fuel turnover, the need for a cheaper chemical processing cycle, and the lack of experience using a liquid sodium circuit.[37] The coolant temperature was designed for 800°F at 127 psi. The steam conditions were to be 740°F, 600 psi.

The original proposal (written in 1957) stated that the Fermi reactor would have a 150 Mw electrical capacity and cost an estimated $45 million, including R&D.[38] The costs of the reactor, including R&D, later rose to $110 million, and the effective plant capacity dropped to 30 Mw.[39] The Fermi plant operated at capacity for only twenty-six days and produced a total revenue of only $.3 million before closing down for lack of funds in early 1973. The total costs of the power plant (not just the reactor) are estimated to be over $130 million (not including $8 million of direct AEC R&D contributions). Of this amount Detroit Edison has paid over $70 million, which exceeds the original total cost estimate of $62 million.[40]

The risks of failure when attempting to employ new technology are often great. Detroit Edison and APDA realized they were endeavoring to design, build, and operate a pioneering venture. Many difficult problems such as redesign and replacement of components added to the project's expense and construction time. Part of the information learned from the Fermi project was that economic nuclear power produced by a commercial fast breeder reactor currently remains an unattained objective of the civilian nuclear power reactor program due to unforeseen difficulties and technical problems. This, of course, does not downgrade the significant contributions of the Fermi project in nuclear power technology. Although the Fermi fast breeder reactor project did not achieve all of its objectives, the power industry has recognized the great value of the knowledge and experience ascertained from the project and the contributions of Detroit Edison and APDA. Much of the information learned from the Fermi experience can be profitably used in future fast breeder reactors.[41]

The first use of an ice condensor containment system at a nuclear power station [60]. A topic of great concern intimately related to nuclear power generation is operational safety. Despite growing criticism of nuclear safety systems' integrity, the AEC states that there is little danger the chain reaction in an operating nuclear power plant can become uncontrollable. The main risk centers on containing the fissionable products created by the nuclear reaction. There are two general types of precautions taken to prevent radioactive material from entering the surrounding environment. (1) The reactor itself is surrounded by a vessel designed to prevent the fuel, coolant, or moderator from escaping the reactor core. This enclosure, the pressure vessel, is made of alloy steels or prestressed concrete, depending on the reactor's operating characteristics. (2) The entire nuclear steam supply system is housed in a containment structure designed to prevent radioactive materials from escaping, even if the pressure vessel fails or leakage occurs outside the reactor. The containment structure is often made of steel or, more recently, of reinforced concrete. The following innovation is one of the more recent improvements in nuclear safety systems.

The first nuclear power plant to use an ice condensor containment system is the Donald C. Cook Plant of AEP's operating division, Indiana & Michigan Electric Company. The first of the two 1,100 Mw units went into commercial service in 1975. If the containment building is to stop fissionable material from entering the environment, it must be able to absorb the tremendous quantities of energy (mostly in the form of heat) released from a failing reactor system. The ice condensor system at the Cook Plant improves the containment's energy absorption capacity by providing a large static heat sink (the ice) inside the structure. The ice, placed around the inside of the containment structure, is capable of rapidly absorbing part or all of the energy released from the nuclear steam supply system. Another benefit derive from using such a system is that the dimensions of the containment structure can be reduced compared to those of a "dry" containment system.[42] This innovation in nuclear safety thereby lowers the total plant capital cost and improves the plant's safety system.

The power generation innovations recounted above were judged to be the most technologically significant of those listed in Table 4.1; they represent slightly over one-quarter of the innovations listed. As expected the Class A & B privately owned electric utilities dominate the list of innovators. Of the five innovations sponsored by publicly owned utilities, three of them were new types of nuclear power reactors built as part of the AEC's Power Demonstration Reactor Program and eventually dismantled because other designs became more feasible.

With only minor exceptions the electric utilities did not participate with the manufacturers in the design or production of this innovative equipment, but rather purchased this new technology from the manufacturers. The risks associated with these utilities' innovations in operation thus are not similar to those faced by the manufacturers, who innovate in design and production. The risks, which arise from being the first company to use the equipment under actual operating conditions, may be significant, since unscheduled outages of newly-designed equipment are far more likely than outages with mature equipment.

Less than ten of the advances listed in Table 4.1 can be classified as capital-saving innovations: examples are the water-cooled turbine [15], single boiler feed pump [10], and supercritical unit [1]. These innovations have decreased the utility's capital cost per unit of output. Innovations designed to reduce the utility's labor cost also are a minority of those found in Table 4.1. Most of these labor-saving innovations ([28] through [34]) not only reduced the required number of plant operators, but improved plant efficiency and realiability as well.

Almost one-half of the innovations undertaken by electric utilities can be classified as capital-using. This evidence may indicate that the record of technological innovation in electric power generation conforms to the Averch-Johnson thesis, although any authoritative pronouncement presumably would await the results of an investigation of technological diffusion of power generation innovations. A development that may accelerate the impact of capital-using innovations is the growing use of nuclear capacity for future power requirements. As shown in Table 4.2, nuclear plants employ relatively more capital than fossil-fueled plants. A utility's rate base would be expanded relatively more if the firm chose to increase capacity with nuclear rather than fossil-fueled units. The installation of nuclear capacity may provide partial empirical support for the existence of the A-J thesis, although the strength of the thesis per se would be difficult to evaluate since utilities presumably have chosen nuclear units for reasons other than their relative impact on the rate base.

The classification of listed innovations by type of activity follows a chronological progression. During the early 1950s many innovations decreased operating costs by increasing thermal efficiencies. Innovations that automated power generation controls were introduced in the late 1950s. Nuclear power generation innovations and pollution abatement techniques, which require more extensive technical resources than past projects, have dominated the technological advances initiated in the 1960s. The trend toward undertaking more costly and time-consuming R&D activities may partly explain why only 18 percent of the innovations listed in Table 4.1 were initiated between 1961 and 1970.

Two additional factors may account for this reduction in innovative output during the 1960s. (1) The rapid growth of aerospace R&D expenditures and corresponding reduction in power engineering graduates from universities contributed to the industry's overall reduction in technical output. (2) The continued increase in demand for power has compelled the electric utilities to place growing emphasis on their power plant's operating reliability. Utilities may have installed new plant capacity of proven, noninnovative design and avoided using newly designed power equipment that would be more susceptible to breakdown and failure.

Having here descriptively analyzed the most technologically significant power generation innovations, chapter V presents an analytical and econometric examination of technological advance in electric power generation.

V

Quantitative Analysis
Of Technological Advance
In Electric Power Generation

The electric power industry's structure of technological advance and the thermal power generation innovations occurring between 1950 and 1970 have been identified, examined, and described in the preceding chapters. This chapter undertakes a quantitative investigation of technological advance in the electric utility industry.

The following microeconomic, empirical investigation of the electric utilities' techno-economic performance is conducted in two segments. First, an analytical examination of the electric utility industry in general and the largest, privately owned electric systems in particular explores the technological performance of the industry's small, medium, and large utilities over the past two decades. The technological performance of the nation's fifty largest systems is then investigated to determine whether the performance of any firm or firms within this group is superior to the others.

Second, an econometric investigation analyzes the largest utilities' technological performance to determine the specific effects of firm size, market characteristics, and R&D expenditures on innovative activity, and the effect of firm size and profitability on the level of a system's R&D expenditures. These two, closely related examinations provide a comprehensive quantitative investigation of the interactions of selected characteristics of the firm and the industry, and of the utility's technological activity.

81

Analytical Investigation of the Electric Utility
Industry's Technological Performance

This investigation examines the 458 electric utility systems that generate 97 percent of the nation's electricity. These systems, which account for less than 15 percent of the nation's operating electric systems, represent the dominant sector of the industry and would benefit the most from advances in power generation techniques.

Table 5.1 arranges the surveyed electric utility systems, which may include more than one individual operating utility, into four size classes. System size is measured by kwh generated rather than total kwh sales to ultimate customers, because the latter measure includes power sold but not generated by the system. Since power purchases from other utilities are sometimes a significant portion of total kwh sales, the use of a sales figure as the measure of system size would tend to inflate artificially the system's size.

Small, publicly owned electric utilities account for two-thirds of the systems in Class 1. In contrast, 90 percent of the utilities in Class 4, the industry's largest firms, are privately owned. The majority of the intermediate-sized systems, those in Classes 2 and 3, also are privately owned, although by a somewhat smaller proportion than the largest class. The average size of systems in Class 4 is five times and two and one-half times that of Classes 2 and 3, respectively.

The structural characteristics of the electric utility industry can be deduced from the figures in columns (c) and (d) in Table 5.1. The twenty systems in Class 4, representing only 4 percent of those systems surveyed, account for almost one-half of the generated output of the 458 systems examined. In contrast, the 385 systems in Class 1, 84 percent of those surveyed, account for only 16 percent of the total generated output of the systems examined. These data illustrate the industry's dominance of a few large, privately owned electric utility systems. The analysis below will determine if the nation's very largest systems also dominate the utility industry's technological performance.

Table 5.1 assesses both the absolute and relative technological performance of each utility class. The innovation ratios, the weighted and unweighted technological output divided by generated output, provide a means of evaluating the class's relative technological performance. The absolute performance is measured by the number of innovations initiated by firms in each class. Both types of technological

Table 5.1. *Electric Utility Systems' Technological Performance by Class*

Utility class*	(a) Number of innovations Unweighted	(b) Weighted†	(c) Number of systems**	(d) Kwh gen.	Innovation ratios (a)/(d)	(b)/(d)
1) Under 5	7 (11)	4.8 (11)	385 (84)	240.2‡ (16)	.03	.02
2) 5 to 9.9	6 (10)	3.9 (9)	31 (7)	210.3 (14)	.03	.02
3) 10 to 20	11 (19)	8.3 (20)	22 (5)	309.2 (21)	.04	.03
4) Over 20	36 (60)	24.8 (60)	20 (4)	719.3 (49)	.05	.03
Total	60	41.8	458	1479.0		

NOTE: Figures in parentheses indicate percentage of column total.

* Class size measured in billions of kwh generated.

† See page 89 for method of weighting innovations by technological importance.

** Number of privately and publicly owned electric utility systems operating in each class in 1970. Publicly owned firms include municipals, state and federal projects. The number of firms in each class is not all-inclusive due to data unavailability, but should be considered comprehensive as these small firms rarely generate their own power requirements.

‡ Total net kwh generated, in billions, by class in 1970. The column represents 97 percent of the industry's generated output in 1970.

83

performance must be examined to discern the technological contribution of these electric utility classes.

The computed innovation ratios presented in Table 5.1 indicate that Class 4, containing the nation's largest systems, has not outperformed the other utility classes.[1] However, Class 4 has outperformed the other three classes in an absolute sense by accounting for 60 percent of the industry's power generation innovations, an impressive statistic. No other utility class has initiated even one-half as many innovations. But it is significant also that when comparing this record relative to the share of industry output of each class, the largest systems' relative technological performance is no better than that of the class containing the industry's smallest firms.

The contribution of the largest electric utility systems, however, cannot be denied. Table 5.2 illustrates the technological impact of the nation's very largest privately and publicly owned electric utility systems. Unlike the four largest coal and petroleum companies' technological performance studied by Mansfield, the nation's four largest electric utility systems (TVA, Southern Company, American Electric Power Company, and Commonwealth Edison Company) did not account for a larger share of the industry's power generation innovations than their share of industry output. Only four of the twelve utility groups listed in Table 5.2 had a larger share of innovations than output.

In absolute numbers, Group 3, in which there were no noninnovators, accounted for the greatest innovative output. Although the innovation ratios of Groups 3, 5, 9, and 12 appear much larger than those of the remaining groups, and despite the wide range of average system size, none of the twelve groups' relative technological performance is significantly different than the others at the .05 confidence limit.[2]

The four largest systems' share of innovative output is not similar to that of the four largest coal and petroleum companies as measured by Mansfield. In these two industries the four largest firms' share of innovations was greater than their share of industry output.[3] Innovative output appears to be more diffused among the largest electric utility companies than in the coal and petroleum industries.

Since the largest electric systems have made a significant contribution to the industry's record of technological advance in power generation, the following analysis examines more closely the technological

Table 5.2. *The Nation's Largest Electric Utility Systems' Innovative Performance*

Utility group*	(a) Number of noninnovating systems	(b) Number of innovations	(c) 1970 total net generations**	(d) Average system size**	(b)/(c) Innovation ratio
(1) 1-4	1	10(17)	253.4(17)	63.4	.04
(2) 5-8	3	4(7)	154.6(10)	38.7	.02
(3) 9-12	0	11(18)	122.4(8)	30.8	.09
(4) 13-16	2	3(5)	103.7(7)	23.4	.03
(5) 17-20	1	8(13)	84.5(6)	21.1	.09
(6) 21-24	2	2(3)	68.6(4)	17.1	.03
(7) 25-28	2	2(3)	61.6(4)	15.1	.03
(8) 29-32	3	1(2)	53.6(3)	13.4	.02
(9) 33-36	2	4(7)	49.0(3)	12.3	.08
(10) 37-40	2	2(3)	41.5(3)	10.4	.05
(11) 41-44	4	0	35.5(2)	8.8	.00
(12) 45-48	2	3(5)	29.9(2)	7.5	.10
Total		50(83)	1,058.3(69)		

NOTE: Figures in parentheses indicate percentage of industry total.

* Systems are ranked by 1970 total net generation; include both privately and publicly owned firms.

** Figures represent billions of kwh.

performance of the nation's fifty largest privately owned electric utility systems.

Almost one-third of the fifty largest privately owned electric utilities are multiple-company systems; altogether they constitute 93 separate operating utilities.[4] These companies, representing less than 2 percent of the nation's operating electric utilities, accounted for 64 percent of the industry's generated output in 1970, for 90 percent of the Class A & B utilities' R&D expenditures reported in 1970, and are responsible for 82 percent of the innovations listed in Table 4.1.

Only twenty-three of these fifty systems have innovated in electric power generation over the past two decades. Seven firms have accounted for 47 percent of the industry's power generation innovations, evidence of the concentration of technological output and of the very small number of technologically oriented firms in the industry. These innovating systems are identified in Table 5.3. Among these firms, the impressive record of the American Electric Power Company is immediately obvious; no other utility has accounted for more than one-half the number of innovations that AEP has pioneered since 1950.

Using the data provided in Table 5.3, the technological performance of each of the nation's fifty largest privately owned electric systems can be examined. The systems with the largest ratio of weighted

Table 5.3. *Innovating Utilities from the Fifty Largest Privately Owned Electric Utility Systems*

Utility	Rank°	No. of innov.	T/G†	T/R&D°°
American Electric Power	2	8	.11	3.3
Commonwealth Edison	3	1	.02	0.4
Southern California Edison	4	4	.06	1.7
Middle South Utilities	8	3	.07	3.4
Detroit Edison	9	3	.06	0.4
Consolidated Edison	10	2	.04	0.9
Public Service Electric and Gas	11	3	.07	2.6
Central & Southwest Corp.	13	2	.08	7.3
Virginia Electric & Power	15	1	.02	0.6
General Public Utilities	16	2	.03	0.4
Allegheny Power System	17	2	.07	4.0
Philadelphia Electric	18	4	.16	3.1
Gulf States Utilities	20	1	.05	2.6
Niagara Mohawk Power Corp.	23	1	.06	1.4
Ohio Valley Electric Corp.	25	1	.06	NA
Pennsylvania Power & Light	26	1	.05	1.4
Wisconsin Electric Power	29	1	.04	2.7
Northern States Power	31	1	.05	0.9
Cleveland Electric Illuminating	32	3	.17	4.6
Long Island Lighting	35	1	.04	0.6
Duquesne Light	37	1	.10	10.7
South Carolina Electric & Gas	43	2	.12	3.0
Public Service of Colorado	45	1	.14	5.5

° Privately owned electric utilities ranked by 1970 kwh generated.

† Ratio represents the number of innovations, weighted by their average technological significance, divided by the firm's generated output in 1970, in billions of kwh.

°° Ratio represents the weighted number of innovations divided by the firm's average R&D expenditures, in millions, incurred between 1966 and 1970.

NA—not available.

innovations to generated output,[5] T/G, is the Cleveland Electric Illuminating Company, the thirty-second largest privately owned electric utility. Although a large firm, it is only one-fifth the size of the Southern Company system, the country's largest privately owned utility system. (Cleveland Electric Illuminating is one-seventh as large as TVA, the largest electric utility in the United States.) When ranked by the size of this ratio, only two of the ten largest privately owned electric systems are included in the ten best-performing electric systems.

The second ratio presented in Table 5.3, T/R&D, measures the firm's technological "output," the number of innovations weighted by their technological significance, relative to its technological "input," the utility's average R&D expenditures over the period 1966–1970. This technological output/input ratio is used as a measure of the firm's technological productivity; when this ratio is large the firm has more efficiently employed its technological resources to produce a technological output.

Once again, the largest systems generally do not rank above relatively smaller electric utility systems. Duquesne Light Company, the thirty-seventh largest privately owned electric utility, has the highest technological output/input ratio, one almost twice as large as any other firm's. Only two of the ten largest firms rank in the ten best-performing utilities with regard to this measure.

The information presented in Table 5.3 suggests that the relative technological performance of the very largest electric systems falls short of that of somewhat smaller systems. Mansfield's analysis of technological performance in the steel industry produced similar results, as the largest steel manufacturers' performance was inferior to that of somewhat smaller steel producers.

The relative technological performance of the class of utilities containing the nation's largest firms was not significantly different from the relative technological performance of the class containing the smallest electric utilities. While absolute technological performance appears directly related to firm size, the industry's very largest systems do not possess nonpareil relative technological performance. The relative technological performance of intermediate-sized firms like Cleveland Electric Illuminating and Duquesne Light ranks much above that of larger firms.

Econometric Investigation of Technological Advance in the Electric Utility Industry

To supplement the descriptive investigation of the preceding chapters and the analytical examination just completed, the following econometric investigation analyzes the process of technological advance in the electric utility industry. The effectiveness of quantitative models such as those presented below can be obstructed by problems inherent in attempts to fully explain as complex a process as a firm's techno-economic activity.

1. The models include only those variables that can be meaningfully quantified. Factors such as regulatory influence, the relative value the firm assigns to the associated prestige of being a successful innovator, the firm's internal R&D structure, and management background all

may play a critical role in determining the success of an electric utility's technological advancement.[6] But these influences could not be meaningfully or reliably quantified and, consequently, are not included in the models.

2. The weights assigned to each firm's innovations are determined subjectively. They measure each innovation's technological rather than economic significance. It should be recognized that despite the precautions taken, any such method of weighting can be subject to bias, and as such, may not accurately reflect the true significance of a particular innovation.

3. The models of innovative activity presented in this study (and those developed by other researchers) examine only successful technological innovation. These models do not undertake any analysis of innovative efforts that were not successful, themselves perhaps an equally important component of innovative activity as technology also may be advanced by recognizing what circumstances may have contributed to the failure of certain innovative ventures.

4. The most serious problem facing quantitative investigations of techno-economic activity is the lack of a resolute theoretical foundation. Although economists have studied the process of technological advance for over forty years, nothing approaching a definitive theory has emerged, especially at the microeconomic level.

The studies mentioned in chapter I have increased the economist's understanding about various aspects of the process of technological advance but have not produced a well-defined theory upon which a precise empirical investigation at the firm level can be based. Of the authors discussed in chapter I, Edwin Mansfield has produced the most complete and comprehensive quantitative analysis of the techno-economic process. But like other researchers, he too has been handicapped by the absence of a well-defined theory explaining the relationship between selected economic variables and technological performance.

With these cautions in mind the following models venture to examine what economic characteristics of innovating electric utilities may account for their technological performance.

Data measuring system size, growth of the system's generated output, its sales market characteristics, profitability, R&D expenditures, and innovative activity were compiled to test several hypotheses regarding electric utilities' technological performance. The technological performance of the electric utility industry's largest companies will be analyzed with particular emphasis placed on testing whether the neo-Schumpeterian hypothesis,[7] relating firm size to R&D expenditures and innovative output, is valid in this industry.

Much of these data was gathered from Federal Power Commission publications. The variables used in the models are presented below.

I/S —The percentage of the system's total kilowatt-hour (kwh) sales sold to industrial customers in 1970.

G_t —The electric utility system's generated output measured by total net kwh generated in year t. This statistic measures the amount of electric power generated by the system's power facilities and does not include kwh purchased by the system from other electric utilities.

$\triangle G$ —The ten-year growth of the system's total net generation, defined as $(G_{70} - G_{60})/G_{70}$.

$R\&D$—Total R&D expenditures reported to the FPC by the system in 1970.

RE_t —The system's rate of return on common equity reported in year t.

T —The number of technological innovations pioneered by the system in thermal-electric power generation from 1950 to 1970, weighted by their technological significance. The weight assigned is the percentage of panel members who judged the innovation to be technologically significant. The members were asked to appraise whether each innovation had or will have a significant impact on electric power generation techniques.

The Fifty Largest Utility Systems' Innovative Performance

The following model examines the relationship between innovative output (T) and system size (G), system growth $(\triangle G)$, market characteristics (I/S), and R&D expenditures $(R\&D)$. To test the validity of the hypothesis that system size and innovative output are directly related requires the introduction of electric utility systems that have not innovated, as well as those that have innovated. The lower limit of the dependent variable (T) is thus zero, with a concentration of observations at this limit. The assumptions of the general multiple regression techniques are invalid, necessitating the use of a more appropriate statistical technique. The TOBIT estimation technique[8] is employed in this model.

The model is defined by the following equation:

$$T = a_0 + a_1 \triangle G + a_2 G + a_3 I/S + a_4 R \& D. \tag{1}$$

The random disturbance term for this and the other equations is omitted from the notation.

The sample includes the fifty largest, privately owned electric utility systems. These systems are often multiple-company systems; they are composed of 93 separate operating utilities. These companies, which represent less than 2 percent of the country's operating electric utilities, accounted for 64 percent of the industry's generated energy output in 1970, for 90 percent of R&D expenditures reported by the FPC in 1970, and are responsible for 82 percent of the innovations listed in Table 4.1. These systems clearly represent a dominant sector of the industry and would benefit most from advances in power generation techniques. Of the fifty systems sampled, twenty-five have innovated in thermal-electric power generation over the past two decades.

Eq. (1) contains the hypothesis that innovative output is directly related to system size (G), percentage of industrial sales (I/S), and R&D expenditures ($R \& D$), and indirectly related to the system growth rate ($\triangle G$).

The first hypothesis (related to the Schumpeterian hypothesis) assumes that only the largest firms, those with significant amounts of available capital resources, will be able to meet the costs and the risks of undertaking technological activity. Since only the nation's fifty largest privately owned electric systems are being examined, the hypothesis states that the largest of these firms will be more likely to innovate and perform technological activity.

It is hypothesized also that electric systems with a high industrial load factor, measured by the percentage of total kwh sales purchased by industrial customers (I/S), will make fuller use of their available base load generating capacity. Systems with a higher proportion of industrial sales are assumed to be more inclined to employ innovative, cost-reducing designs in their new generating equipment, since industrial customers' demand for electricity appears to be relatively price elastic and thus may be significantly affected by price changes in the long run.[9]

The weighted number of innovations initiated by the system is hypothesized to be directly related to the level of its R&D expenditures. The greater the system's technological input, the more technologically oriented it is, thus the more likely the system is to have innovated.

Finally, electric utility systems with lower rates of growth in generated output are assumed to be more likely to be motivated to innovate and introduce innovative, cost-reducing generating capacity. Since construction of new generating facilities takes a sizeable amount of time, a ten-year period was selected as the appropriate time span to measure the growth of generated output.

The parameter estimates, with the t-ratios appearing in parentheses given below, confirm only two of these hypothesis.

$$T = -.774527 - 1.77920 \triangle G + .04120 \ G \qquad [1]$$
$$(1.67)^* \qquad\qquad (2.42)^{**}$$
$$-2.11064 \ I/S + .046369 \ R\&D.$$
$$(1.55) \qquad\qquad (.23)$$

* Significant at the .05 confidence limit.   **Significant at the .01 confidence limit.

The hypothesis that innovative output and system size are directly related is supported by the data. The larger the system becomes, the more able it is to meet the costs and risks of undertaking technological innovation. Since all the innovations listed in Table 4.1 are capital-embodied, it is not unexpected that the larger systems, those with more technical and economic resources available to them, would have innovated more frequently. More will be said about the relationship between technological activity and system size in the models that follow. On its face, therefore, the record of technological innovation appears to conform to the neo-Schumpeterian hypothesis.

Also supported is the hypothesis that innovative output and system growth are indirectly related. Electric utility systems with relatively low rates of growth in their generated output appear to have greater incentive to spur future load growth by employing innovations that would realize lower unit generating costs.

The data do not support the hypothesis that systems with proportionately greater industrial sales nor those with larger R&D expenditures have been more likely to innovate. The most probable reason for this latter result is that in many cases the R&D expenditures undertaken by the electric utilities are only peripherally related to the types of power generation innovations listed in Table 4.1. The R&D input for these innovations has been heavily financed by the electrical equipment manufacturers, not by the utilities. While the organization of industry R&D activities through the Electric Power Research Institute may serve to ameliorate the utilities' R&D performance, during the period analyzed the utility's R&D activities had a nonsignificant effect in explaining innovative performance.

Innovating Utilities' Technological Activity

The following model examines in more detail the technological activity of those privately owned electric utilities that have innovated in thermal-electric power generation. Technological activity is here measured by the system's record of power generation innovations and its R&D expenditures. The model is defined by two simultaneous equations,

$$T = \beta_0 + \beta_1 \triangle G + \beta_2 G_{70} + \beta_3 I/S + \beta_4 R\&D. \tag{2}$$
$$R\&D = \gamma_0 + \gamma_1 G_{67} + \gamma_2 RE_{67} + \gamma_3 T. \tag{3}$$

The hypotheses defined in Eq. (2) are similar to those of Eq. (1). Eq. (3) contains hypotheses that R&D expenditures are positively related to the system's past size (G_{67}), profitability (RE_{67}), and innovative output (T). It was determined from interviews with electric power industry personnel that between one and five years may pass before R&D projects undertaken by electric utility systems are completed. From this information, profit rates, measured by the rate of return on common equity (RE) and subject to regulatory commission scrutiny, and system size (G) were lagged three years to measure their effects on R&D projects of average duration.

The inclusion of innovation (T) into Eq. (3) defines the potential feedback effect: that systems having undertaken successful innovative activity in the past will have more incentive to continue their technological inquiries in the form of ongoing R&D projects.

Eqs. (2) and (3), defining the hypothesized relationships between an electric utility system's economic characteristics and its technological activity, are the structural equations in a system of two simultaneous equations. The Full-Information Maximum Likelihood regression technique, employing Broyden's Rank-One-Correction Method,[10] was employed to estimate the equations' parameters. Variables R&D and T are the endogenously determined variables of Eqs. (2) and (3), respectively. The sample includes the twenty-nine innovating privately owned electric utility systems.

The parameter estimates for Eqs. (2) and (3) are presented below.

$$T = -1.02739 - 1.15731 \; \triangle G + .083719 \; G_{70} \qquad [2]$$
$$(1.43) \qquad\qquad (3.06)^{**}$$
$$+ \; 4.21483 \; I/S - .337616 \; R\&D; \; R^2 = .89; \; DW = 2.22.$$
$$(3.06)^{**} \qquad (.89)$$
$$R\&D = .009999 + .076095 \; G_{67} - .027154 \; RE_{67} \qquad [3]$$
$$(4.07)^{**} \qquad\qquad (.49)$$
$$+ \; .032304 \; T; \; R^2 = .89; \; DW = 2.37.$$
$$(.13)$$

** Significant at the .01 confidence limit.

The parameter estimates of Eq. (2) support two of the hypotheses, that system size and industrial load are directly and significantly related to the number of innovations initiated by the utilities. The remaining hypotheses concerning system growth and R&D expenditures are unsubstantiated by the data.

Surprisingly, innovation and R&D are only weakly correlated ($r = .38$). The most probable reason for this is again that in many cases the R&D expenditures undertaken by the electric utilities are only indirectly related to the types of power generation innovations listed in Table 4.1.

If the record of power generation innovation over the past two decades is a guide (and there is no reason to believe otherwise), the largest electric utility systems would have to reorient their R&D programs from striving to produce better operating systems to producing more advanced electric power equipment in order to effectuate innovations similar to those listed in Table 4.1. Such reorientation has begun to occur recently within the R&D activities of industry associations, but it has yet to have a strong impact on the utilities' technological output. In summary, the lack of statistical significance of the R&D variable in Eq. (2) serves to document the minor impact that utility R&D has had on power generation equipment innovation.

Although these systems' R&D expenditures have not significantly influenced innovative output, the parameter estimates of Eqs. (2) and (3) show that firm size is related to the utilities' technological activity. When examining those systems that have innovated, we find that both innovative output and R&D expenditures are directly related to system size.

It is of some interest to find that the estimates in Eq. (3) show that the electric utility system's profitability (RE) does not significantly affect the level of its R&D expenditures.

Assuming that regulatory control exerts a meaningful and observable influence on the utilities' rate of return, this result may substantiate the contentions of those utility industry spokesmen, such as Donald C. Cook and Philip Sporn mentioned on page 23, that regulation, contrary to some people's preconceived view, has not acted as an impediment to the utilities' R&D activity.

The lack of statistical significance of the profitability variable in Eq. (3) may be explained by the method in which many electric utility managements engage in R&D ventures. Interviews with utility personnel revealed that few, if any, managers explicitly take into account profit rates when deciding what may influence the level of the company's R&D activity.

Finally, the lack of statistical significance of innovation (T) in Eq. (2) provides some evidence that the hypothesized technological feedback is not present among the largest privately owned electric utility systems. The past record of successful technological innovation appears not to affect systematically the present level of the firm's R&D expenditures. A more appropriate measure of this potential feedback might be the ratio of successful to unsuccessful innovations. This ratio, however, was not employed, since data for unsuccessful innovations were not available.

R&D Expenditures and Utility Size

To test the important question of whether the largest privately owned electric utility systems spend more on R&D relative to their size than smaller systems, Eq. (4) assumes a log-linear form. It is hypothesized that the largest electric systems spend more on R&D relative to their generated output (G) than smaller firms, thus the elasticity of R&D expenditures with respect to the firm's generated output, δ_1, is assumed to be greater than one.

Using Mansfield's studies of the chemical, steel, petroleum, drug, and glass industries as a guide, the analysis is now extended to the electric utility industry. While Mansfield obtained results from sampling only four glass companies and eight drug manufacturers, the forty-nine largest private electric utility systems' R&D expenditures in the years 1970 through 1973 are examined in Eq. (4).[11] The stated hypothesis is based on an argument espoused by many persons inside the electric utility industry and is similar to that argued by Schumpeter, Villard, and Galbraith.[12] The costs of maintaining a successful R&D program are large and growing larger as time passes, so that the larger firms are more able to engage in R&D than smaller firms. The hypothesis is defined in the equation

$$lnR\&D_t = \delta_0 + \delta_1 \ln G_{t-3},\qquad(4)$$

where t varies from 1970 to 1973. The estimates, determined by ordinary least squares, are given in Table 5.4.

Table 5.4. *Parameter Estimates for Equation (4)*

Data set	Year	δ_0	δ_1	F	R^2	Number of observations
A	1970	−3.87122 (8.25)	1.23185 (6.83)**	46.59	.50	49
B	1970	− .05102 (.14)	.7186 (8.68)**	75.36	.49	81
A	1971	−4.53398 (1.24)	1.10084 (4.97)**	24.66	.34	49
B	1971	−1.39438 (1.11)	.89769 (11.20)**	123.42	.62	80
A	1972	−4.87258 (1.25)	1.15391 (4.88)**	23.82	.34	49
B	1972	−1.56774 (1.26)	.94453 (11.97)**	143.17	.65	80
A	1973†	−3.40285 (1.01)	1.08950 (5.37)**	28.85	.38	49
B	1973†	2.16538 (1.76)	.73752 (9.48)**	89.89	.54	80

NOTE: t-statistics appear in parentheses.
** Significant at .01 confidence limit.
† 1973 data are taken from *Congressional Record*, 19 December 1971.

The parameter estimates for Eq. (4) were derived using two related data sets, labeled A and B in Table 5.4. Data Set A refers to those data measuring the net generated output and R&D expenditures of the forty-nine largest utility systems, measured as systems; e.g., American Electric Power Company is accounted for as one system. Data Set B measures the same forty-nine utility systems but disaggregates these systems into their component operating utilities; e.g., the units of observation are the operating utilities that reported R&D expenditures such as Indiana & Michigan Electric Company, Kentucky Power Company, Ohio Power Company, Appalachian Power Company, and Michigan Power Company, which compose American Electric Power Company.

Let us first examine the results using Data Set A. The parameter estimates show the size elasticity of R&D expenditures, δ_1, not to be significantly different from unity.[13] Thus from 1970 to 1973 there was no significant difference between the R&D expenditures reported by the industry's largest systems relative to their size and those of somewhat smaller ones, which of course are still large companies by most standards. This constancy of the size elasticity of R&D expenditures is rather surprising, since it was measured over a period of rather intense external and internal pressure for the electric utilities to increase their spending on R&D. These results place the sampled electric utility systems with firms in the steel industry, where Mansfield found that the largest steel firms did not spend more on R&D, relative to their sales, than did smaller firms.[14]

Now let us investigate the results when Data Set B is used to regress R&D expenditures on firm size, now measured by operating utilities' generating output. The results in Table 5.4 show that the size elasticity of R&D for the years 1970 and 1973 was significantly below unity. But for 1971 and 1972, δ_1 was not different from unity.

Two questions arise: first, why in 1970 and 1973 do the estimates of δ_1 change significantly between using Data Sets A and B; and, second, what may be the reason for the shift back and forth of the parameter estimate over time?

The first question of how to explain this change in the size elasticity of R&D between using both Data Sets may be answered by examining the impact of the companies' managerial efficiency and the institutional and/or economic factors that may influence how R&D funds are allocated among the operating utilities in an electric utility system. For example, Northern States Power Company, comprised of two distinct operating utilities, Northern States Power Company, Minnesota and Northern States Power Company, Wisconsin, has exhibited the type of allocation of R&D expenditures between operating companies that could cause changes in the δ_1 when comparisons are made between system and operating utility data. Table 5.5 shows that between 1970 and 1973 the total R&D expenditures for Northern States

Power more than doubled. But the distribution of R&D between its Wisconsin and Minnesota utilities completely reversed. In 1970 the Wisconsin utility accounted for just over one percent of the system's total R&D expenditures. By 1973 the Wisconsin company accounted for over 95 percent of the system's R&D. Such internal reallocation of R&D funds can likely cause significant changes in the firm versus system parameter estimates.

Table 5.5. *R&D Expenditures of Northern States Power Company and Its Operating Utilities, 1970–1973*

Firm	1970	1971	1972	1973
Northern States Power Co. (Wisconsin)	$ 14,241	25,272	97,586	2,558,276
Northern States Power Co. (Minnesota)	$1,119,579	1,401,108	1,883,564	128,618
Total	$1,133,820	1,426,380	1,981,150	2,686,894

SOURCE: Federal Power Commission, *Statistics of Privately Owned Electric Utilities in the United States,* years indicated.

Again, one would expect *a priori* that the effect of factors motivating the utilities to spend relatively more on R&D over this time frame would be visible. Instead, by 1973, sampled by systems, the estimated coefficient is not significantly greater than unity. When the electric utilities are sampled as individual operating companies the estimated elasticity coefficient is significantly less than unity.

The second question, regarding sensitivity of the parameter estimates over time, is potentially a more difficult one. One possible problem lies with the assumption that only net generated output can explain changes in the level of R&D expenditures. This is, of course, a simplifying assumption that can be relaxed. When variables measuring profitability of the utility were added, similar to Eq. (3), they were not statistically significant. Until more recent data become available, the question of structural stability of Eq. (4) will remain.

This analysis does not presume that all electric utilities are able to or choose to spend money on R&D; an as-of-yet undefined minimum firm size is probably required before an electric utility feasibly can sponsor R&D activities. Also important is that over the time period examined, the utility industry has increased markedly its R&D expenditures in an absolute sense. From 1970 to 1973 the reported R&D expenditures rose over 500 percent to $239,223,194. But the above results indicate that relatively smaller utility systems have been able to make meaningful contributions to the total electric utility industry's R&D efforts on a par with the largest systems.

The "Optimal-Sized" Utility for Thermal-Electric Power Generation Innovation

Events such as the establishment of larger, more sophisticated power pools and the creation of more centralized, cooperative R&D activities through the Electric Power Research Institute are likely to have a substantive impact on the utility industry's future dynamic performance. Before the full impact of such developments on the industry's conduct and operations can be assessed, the relationship between innovative output and firm size should be clearly understood. The following analysis was conducted to determine the "optimal-sized" utility with regard to innovative output in thermal-electric power generation. Assuming that *all* other factors except the utility size exert an insignificant effect on innovative output, some preliminary answers can be obtained.

The Box-Cox transformation was applied to Eq. (5) to specify the appropriate functional form for the relationship between utility system size (G) and innovative output (T).[15]

$$(T^{\lambda_0} - 1)/\lambda_0 = \psi_0 + \psi_1 [(G_1^{\lambda_1} - 1)/\lambda_1]$$
$$+ \psi_2 [(G_2^{\lambda_2} - 1)/\lambda_2], \tag{5}$$

where

$$G_2 = G_1^2.$$

The sample includes the twenty-nine innovating privately owned electric utility systems. The results, given below, suggest that the relation in Eq. (5) is linear in the logarithms of the variables since λ is not significantly different from zero for any parameter estimates. Given this information from the Box-Cox transformation, the appro-

$$(T^{-.235511}_{(.87)} - 1)/.235511 = -0.514287 - 0.0355710 \tag{5}$$
$$(G_1^{-6.16047}_{(.01)} - 1)/-6.16047 + 0.015555 (G_2^{.479097}_{(1.31)} - 1)/$$
$$.479097;$$
$$R^2 = .36; \; DW = 2.40$$
$$\overline{\chi^2 = 14.41}.$$

priate functional relationship is defined in Eq. (6). The parameter estimates for Eq. (6) were obtained using the ordinary least-squares

$$\log T = \zeta_0 + \zeta_1 \log G + \zeta_2 \log G^2 \tag{6}$$

regression technique and are given below.

$$\log T = -0.591475 + 87.2584 \log G - 43.3708 \log G^2; \tag{6}$$
$$(1.85)^* \qquad\qquad (1.85)^*$$
$$R^2 = .33; \; DW = 2.28$$
$$\overline{F = 6.41}$$

* Significant at the .05 confidence limit.

While caution should be observed in making any final conclusions, it is of interest to note that the number of innovations reaches a maximum in Eq. (6) when total net generated output (G) equals 10.14 $\times$ 10^9 kwh. Thus from this sample of innovations, the relatively smaller utility systems appear to be capable of significant technological performance. This dynamically "optimal" system size does not occur over the very largest systems, but is representative of firms such as Cleveland Electric Illuminating, and Duquesne Light. In 1970 Dusquesne Light generated 10.32 $\times$ 10^9 kwh and was the nation's thirty-seventh largest privately owned electric utility system. Once again, this analysis of the electric utility industry's dynamic performance is shown to be similar to Mansfield's investigation of the steel industry, where the data also did not fully support the neo-Schumpeterian hypothesis.[16]

The econometric models of the electric utilities' technological performance just described have presented preliminary evidence that large electric utility systems and those with a higher proportion of industrial sales are more likely to have innovated in electric power generation and that the system's R&D expenditures have exerted an insignificant impact on its technological innovation. As would be expected, system size is directly related to the firm's R&D expenditures. While it would be faulty to conclude from these results that by increasing its industrial load or its generated output a utility will then innovate, such characteristics as size and industrial load appear to contribute to the economic environment necessary to make technological innovation both possible and feasible. From the parameter estimates of Eq. (3)

system size also appears to be a necessary ingredient to enable utilities to allocate larger amounts of funds in R&D activities, although probably still small sums relative to the system's size. Since the largest systems do not spend relatively more on R&D than somewhat smaller utilities, it appears correct to infer that the performance of the electric utility industry's largest systems has not been technologically superior to that of relatively smaller firms.

More analysis should be completed before a definitive answer can be given regarding the relationship between utility system size and overall technological performance. However, these preliminary results have some important implications about the industry's dynamic performance. They have shown that there is a positive and significant relationship between the number of thermal-electric power generation innovations and system size; that the very largest systems have not spent relatively more on R&D expenditures than somewhat smaller systems; and that the number of such innovations is maximized over relatively small systems. Further evidence should be presented before an authoritative determination about overall technological performance can be made confidently; for now the results of the models shown above strongly suggest that while the larger utility systems have excelled in absolute terms, their relative technological performance is at best only equal to that of smaller electric utilities.

VI

Conclusions and Recommendations

Two events emerging during the last decade have seriously strained the existing technical resources of the electric power industry: the continually increasing demand for electric power and the public clamor for a cleaner environment. Both have created a pressing need for expeditious, exhaustive solutions to technological problems whose resolution will require development of techniques for generating electric power that are efficient, economical, and environmentally safe.

Historically, the vast majority of electric utilities have chosen not to pursue actively pioneering technological activity. This choice was easily made and sensible, from the utilities' viewpoint, as long as the equipment manufacturers continued to provide improved machinery of increasing size and efficiency and as long as regulatory commissions did not encourage the utilities' participation in technological activities. Only recently have the electric utilities begun to participate more actively with the equipment manufacturers and the government to solve the industry's technical problems. Although its technological input may increase as a result of activities of the Electric Power Research Institute and other groups, it is unlikely that the fundamental structure of technological activity in the power industry will change significantly. What is certain is that more money will be spent by all sectors on power-related R&D in the coming years due to changing social, economic, and technical priorities.

Much of this increased technological activity will involve further nuclear and nonnuclear R&D projects such as the construction of a commercial fast breeder reactor, critical examination of nuclear reactor safety systems, solving the long-run containment problems of spent nuclear fuel, development of "renewable" energy systems such as solar, geothermal and tidal energy, and creating commercially operable and effective pollution abatement equipment for existing nuclear and fossil-fueled plants.

Principal Findings Summarized

The principal findings of this investigation are summarized as follows.

1. The electric utility industry, one of the nation's largest, has been building new electric plants of ever-increasing size since 1950. But over the past decade, its diminishing yearly gains in generating efficiency (shown in Figure 2.2) point to the important technological barriers that remain unsolved.

Despite the growing need, the electric utility industry's relative R&D expenditures are only one-tenth of the 1969 all-industry average of R&D/net sales. Of the more than three thousand operating electric utilities in the United States, the fifty largest privately owned systems, generating 64 percent of the nation's electricity, have accounted for 90 percent of the R&D expenditures reported by the FPC in 1970 for Class A & B utilities.

2. The utilities' technological performance has been affected largely by the industry's structure and by managements' strategies in purchasing new generating capacity. Electric utilities that only distribute electric power, about 70 percent of the nation's operating utilities, and those that generate power from hydroelectric sources have had little incentive or cause to invest in fossil-fueled or nuclear power generation R&D. When faced with the sometimes conflicting goals of pursuing technological advances in electric power generation and maintaining reliable service, many utility managements have sought the expediency of purchasing equipment of proven, noninnovative design. Other managements have pursued a strategy of realizing production scale economies rather than installing equipment which employs higher thermal efficiencies and utilizes new technology. In general, utility managements have adopted a policy of technological followership by assuming that others, be they utilities, manufacturers, or the government, would initiate and undertake the industry's needed technological activity.

The presence of state and federal regulation, on balance, appears to have had little impact. Apart from the FPC's 1970 R&D accounting changes designed to provide some financial inducement, regulation has done little to encourage the utilities to undertake technological activities that could culminate in innovation and greater system efficiency.

3. Five groups within the electric power industry perform its technological activity: the federal government; the equipment manufacturers; the electric utilities, individually or in groups; industry associations; and universities. Figure 6.1 illustrates the tendency toward specialization in technological activity by these groups over the past two decades. Government-financed R&D, usually directed toward basic research, has dominated technological activity in nuclear power generation. Generally, both the manufacturers' and utilities' technological activities are more commercially oriented than those of the government. The equipment manufacturers have concentrated their considerable technological resources on equipment design, development, and improvement. Most utility R&D has been concentrated

Figure 6.1. *Input by Sector in Electric Power Industry Technological Activity*

Percentage of total activity

100

More basic, theoretically oriented activity

More applied, commercially oriented activity

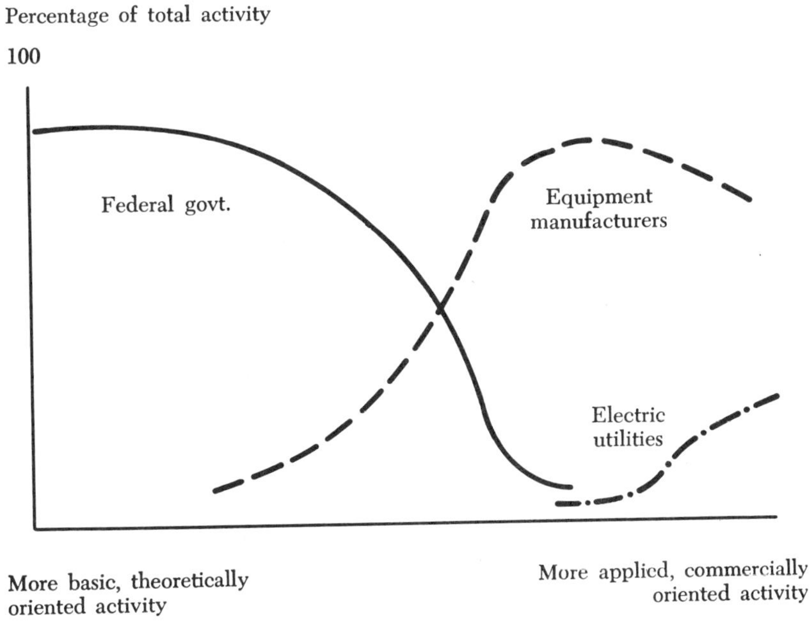

on applied research, examining relatively short-term, system operating problems. Recently the utilities and their industry associations, reacting to both external and internal pressures, have begun to exercise a greater role in sponsoring industry R&D, although they are still vastly overshadowed by the government's and the manufacturers' efforts. The motives associated with utilities' technological efforts, all highly intercorrelated, include the desire to compete effectively with other energy producers, to increase the demand for electric power, to reduce operating costs, to increase service reliability, and to promote personal and corporate prestige.

4. The utilities innovated by purchasing technological advances as they were offered from the equipment manufacturers but, with few exceptions, did not participate in the innovation's design and development. Whereas the manufacturers incur the risks of design and production, the innovating electric utilities take the risks associated with being the first company to use the equipment under actual operating conditions. These risks may be substantial, since newly designed equipment is more prone to failure than equipment of proven design.

Each of the sixty electric power generation technological innovations identified in Table 4.1 is capital-embodied. They have increased thermal efficiency, automated plant controls and operation, reduced the plant's environmental impact, or introduced advances in nuclear power generation. The observable capital-using bias of these innovations lends credence to the application of the Averch-Johnson thesis to technological advance in electric power generation. Innovative output in the 1960s decreased significantly from that in the 1950s due in part to the rapid growth of aerospace R&D expenditures, a reduction in university-trained power engineers, the increased emphasis placed on service reliability, and the lengthening lead times needed to develop new generation technology.

5. The industry's largest systems do not significantly outperform the small and intermediate-sized ones in terms of relative technological performance. The largest systems, however, dominate in absolute innovative output. The fifty largest privately owned electric utility systems have accounted for 82 percent of the power generation innovations discovered between 1950 and 1970. Technological output is concentrated among relatively few utilities within this group. Seven firms have accounted for 47 percent of the industry's power generation innovations.

System size and industrial load were found to be directly related to innovative output, whereas R&D expenditures and system growth did not significantly explain the system's innovative output. System R&D expenditures were found to be related directly to past system size, but were not statistically significantly affected by system profitability and innovative output. It also was determined that among the nation's largest systems, the biggest firms did not spend more on R&D, relative to their generated output, than somewhat smaller systems.

Unexplored Areas

This monograph has been confined to the process of technological advance in steam-electric and nuclear power generation, which is but one facet of the process of technological advance in the energy sector. Further exploration would prove fruitful in the following areas.

The important role of the equipment manufacturers in pioneering technological advance has been documented here. A study of selected power generation innovations initiated by them, complementing the research done here, should lead to a more thorough understanding of what criteria have been applied in choosing R&D projects, how decisions are made to undertake innovations in production, and how this innovative equipment came to be purchased by the utilities.

As part of a more detailed investigation of the decision-making process associated with the electric utilities' technological activity, answers might be obtained to such questions as: Do the electric utilities compute an expected rate of return on each R&D project before they commit any funds to it, and if so, how is such a return computed? Does the regulatory process—which now allows most, if not all, R&D expenditures into the utilities' rate bases where they are paid by the firms' customers—make it unnecessary to compute such individual project returns? What economic, technical, or social standards are applied in making the decision to undertake a particular R&D project? What criteria are employed to allocate R&D expenditures among the operating divisions of multicompany systems?

Another avenue of investigation opened by the results of this study would be to investigate whether the electric utilities differ in their estimate of the commercial usefulness of accessible new technology and compete among themselves to be the first to use it. Certain utilities have concentrated on employing specific types of technology: TVA apparently has specialized in purchasing larger-sized steam generating units; Commonwealth Edison has made a sizeable commitment in nuclear technology for steam generating plants; other firms have adopted new technology employing improved operating temperatures and pressures. It is important to determine the extent of such dynamic competition which is essential to the long-run vitality of individual utilities and the entire industry: diversity of innovative technique and technological competition would encourage continued economic advancement in the industry by keeping narrow the gap between the industry's record of achieved progress and the potential, but unknowable, maximum rate of economic and technical progress.

A logical sequel to this study would be an examination of the process of technological diffusion in the electric utility industry which would identify those economic characteristics common to utilities that have chosen to employ recently developed technology. For example, it might be determined that installed equipment cost and fuel costs are significant factors in explaining which firms have installed supercritical pressure and reheat units.

An equally important area of research is that of determining how the regulatory framework may affect a utility's choice of generating technology, possibly further identifying a cause-and-effect relationship between regulation and the capital-using bias of generation innovations. The influence, if any, of regulation on the utilities' marginal rate of technical substitution also could be examined. In addition, a more detailed study of regulation's effect on technological advance could provide recommendations on what type of institutional structure would maximize the utilities' dynamic performance by promoting innovation and technological competition, and by optimizing the rate of technological diffusion.

Finally, a potentially important relationship was observed between factor price increases and technological innovation and diffusion. As input prices steadily rise, there is, of course, more incentive to employ factor-saving technology. Table 2.6 has documented the rapid increase in the real fuel cost faced by electric utilities. Because fuel costs have risen, coal gasification technology developed over the past decade may now be economically practical. Thus these fuel price increases may have made existing experimental technology commercially feasible. Future research could be undertaken to determine if existing "cost-push" technology, introduced as a result of changes in factor costs, can be distinguished from newly developed "demand-pull" technology undertaken because of the need to expand productive output by using more efficient production techniques.

The research suggested above illustrates that this study is but an initial contribution to understanding the process of technological advance in the electric utility industry. Such additional research could render a valuable social service not only by leading to a more complete understanding of the process of technological advance, but also by improving the technological performance of the electric utilities.

Appendix A

A Brief History of Public Regulation
In the Electric Utility Industry

The influence of regulation on the electric utilities has been examined in this study only as it may affect the firm's ability and incentive to perform technological activity. An examination of the general history and intent of regulation is included here since public regulation exerts an undeniable influence over this industry's economic and technological activity.[1]

Various goals affecting the performance and characteristics of economic activity are incorporated in any system. These goals may include: (1) continued technical efficiency to keep production costs as low as possible, (2) allocational efficiency, and (3) distributional equity. In the United States the competitive system has been relied upon most frequently to achieve these goals. But unrestricted competition may not always best serve the public's interest. When the decision to publicly regulate a firm or group of firms is made, competition is at least partially supplanted by public control. Regulation must then protect the public's interest with regard to the industry and its economic performance.[2]

A business activity needs to be classified as a public utility before public regulation of prices and profits is usually considered. The evolution of the notion of a public utility came to emphasize two ideas: those of monopoly and of necessity.

Historically, government regulation of business was present in the Greek and Roman empires. The doctrine of "just price" (by some interpretations, the price that will repay the expenses of production) dates from Aristotle, Saint Augustine, and Saint Thomas Aquinas. Highways, wharves, business terminals, and water companies long have been viewed as public utilities and placed under public regulation. In the United States, judicial consideration of public utility regulation began with the Supreme Court decision in *Munn v. Illinois* [94 U.S. 113 (1877)]. This decision involved the validity of an Illinois statute setting maximum prices for storing grain in elevators at Chicago under virtually monopolistic market conditions. The Court's opinion formed the foundations for modern utility regulation, finding that the states did have a right to set prices for certain types of business activity. Chief Justice Waite stated in his decision that when a person submits his property to a use in which the public has an interest, he submits to public control to the extent of that interest.

As monopolists selling a product that has become a necessity to many people, the producers of electricity rightfully can be considered public utilities. The industry's regional monopolization was partially induced by the techno-economic nature of electric power generation and transmission: unit production costs diminish as output increases; so-called wasteful duplication of transmission and distribution facilities exists when there is more than one local distributor of electricity; and truly long distance transmission was, until recently, economically infeasible due to technological constraints limiting transmission voltage levels.

As the number of residential, commercial, and industrial uses for electric power increased, electricity became more of a necessity. Electricity was first commercially produced to compete with gas and oil for the home lighting market. Compared to gas or oil, electric lighting became recognized as the superior method of home lighting since there was no danger of fuel explosions or the problem of soot deposits in the home. The demand for electricity grew with the public relying on it for more varied uses. As the public's need increased, the economic conditions that gave rise to the industry's regional monopolization also intensified. Electric companies, having become monopolists selling a necessity, assumed the characteristics of public utilities. As public utilities they became subject to public regulation.

Public utility regulation by state commissions began after the weaknesses and inefficiencies of local regulation became recognized. The first, modern state regulatory commissions date from 1907, when New York, Wisconsin, and Georgia enacted legislation forming permanent state regulatory commissions. Other states soon established public utility commissions; by 1917, twenty-four states had created state regulatory commissions. The principal powers accorded these commissions remain: service, financial, and rate regulation; the prescription of state utility accounting systems; and the issuance of certificates authorizing services and expansion and abandonment of facilities.

Federal regulation of interstate electric utility operations began with the passage of the Federal Water Power Act of 1920 creating the Federal Power Commission. At the time, federal regulation was limited to the licensing of applicants seeking to construct hydroelectric power projects on the navigable waters of the United States. The Federal Power Act of 1935 and the Natural Gas Act of 1938 expanded the FPC's regulatory powers to include the transmission and sale at wholesale rates of electric energy in interstate commerce, and the transportation and sale for resale of natural gas in interstate commerce.

The aims of public utility regulation are consistent with the general economic goals mentioned above. A major purpose of public regulation of the electric utilities is to prevent economic exploitation of customers. Traditionally, regulation has been most concerned with the level of prices charged to the electric utility's customers and with the rate of return paid to the utility's investors. The regulatory procedure for evaluating the utility's prices and rate of return is described below.

By using the formula

$$R = E + (V - D)r$$

[where R = total operating revenue; E = operating expenses; V = fair value of the firm's property, when new; D = accumulated depreciation allowance; and r = rate of return on the depreciated value of property; $(V - D)$ = the rate base, upon which the utility earns a return], the regulatory commission determines what a "fair value" and a "reasonable" rate of return should be for the utility. Without elaborate detail, the evaluation process usually proceeds as follows.

The commission determines what the utility's operating expenses are, including salaries, pensions, all business-related taxes, sales expenses, and production, transmission, and distribution expenses. The utility's rate base is independently evaluated using various accounting techniques for determining the value of plant, property, equipment, an allowance for working capital, and accumulated depreciation.

Given information about recent economic conditions, the regulators determine what the allowable rate of return, after taxes, should be. The utility's past performance, based on a test year, is examined to facilitate the commission's decision of what should and should not be allowed as part of operating expenses, the rate base, and the rate of return. The decision to allow or exclude certain items from the firm's operating expenses or its rate base can exert a pronounced influence on the amount of the firm's business expenditures such as R&D activity.

The price of electricity to be charged by the electric company is determined after the firm's operating expenses and the rate of return are known. Using extrapolations of recent sales volumes, the final unknown (the price of the product) is found that will equate total revenue with total costs.

This method of public utility regulation has been unaltered for decades. The question of regulation's effectiveness has been examined elsewhere[3] and will not be discussed here.

Appendix B

The Process of Steam-Electric
Power Generation

The following simplified description of power generation by a typical fossil-fueled steam-electric power plant[1] should give the reader unschooled in power engineering an understanding of the changes in power generation that have taken place since 1950.

Electric power is generated by tapping the energy contained in the fuel to produce electric energy. Basically, water heated by burning fuel changes to steam. The steam, under pressure, moves a series of turbine blades, driving the generator, which produces electric current. The major pieces of equipment used in this process are the boiler (steam generator), the turbine, and the generator.

The largest single piece of equipment in a power plant is the boiler, often five stories tall and sometimes producing over six million pounds of steam per hour. Most boilers consist of an enclosed series of tubes connected at either end to a steel drum. These tubes are surrounded by air heated to high temperatures by the burning fuel underneath. Almost all modern fuel systems use either gas, oil, pulverized coal, or some combination of fuels. Pulverized coal[2] is blown into the furnace, mixed with air, and ignited; it burns much like a gas flame, producing a fireball in the furnace. In modern boilers the fireball's position can be vertically adjusted by movement of the burners to regulate the temperature of the boiler gases. The heat of these gases is absorbed by the tubes and the water inside them. As the water rises through the tubes to the upper drum, called the steam drum, it changes to steam. There steam collects to be piped into the turbine.

The turbine changes the thermal energy of the steam into kinetic energy by using the steam to rotate a shaft connected to the generator. The turbine consists of series of blades, each radiating from a common shaft.[3] The steam, directed under pressure through a series of nozzles, pushes against the first series of blades, moving the shaft. In the process, the steam loses some of its energy and, consequently, some of its pressure and temperature. At the lower pressure, the steam turns the next series (stage) of blades. The

111

lower steam pressure necessitates longer turbine blades (more surface area) to produce the same thrust. The stage with the longest blades is the last (lowest pressure) stage. In 1950, the last stage blades were twenty-three inches long. By 1967, the blade length had increased to thirty-one inches, which provided greater efficiency.[4]

The steam leaves the last stage of the turbine and is condensed back to water. As some leakage of steam occurs in passing through the steam generator and turbine, raw water, or make-up as it is called, is placed into the system. The make-up is preheated before entering the boiler by being piped through the exhaust steam in a feedback heater. The boiler feedwater (composed of make-up and condensate) is further heated by being pumped through an economizer, a series of tubes inside the boiler around which pass flue gases. Although much of the heat of these gases has been absorbed by the boiler tubes, enough remains to raise the temperature of the feedwater close to the temperature of the water in the boiler proper.

One of the critical pieces of equipment in the steam generation cycle is the boiler feed pump, which moves the feedwater from the condensor through the economizer to the boiler drum. Unless this pump is continuously operating, the steam generator may run dry within minutes, causing it to rupture. To prevent such an occurrence, there are usually a series of boiler feed pumps, each operating at pressure high enough to overcome the boiler drum pressure.

The electric generator changes mechanical energy into electric energy. It is composed of a rotating magnet connected to the turbine shaft and surrounded by stationary coils of wires. Electric current is produced in the stationary coils as they interfere with the moving magnetic field. The modern generator is a very efficient machine; some have operating efficiences of 97 percent. Improvements in generator design have centered mainly on increasing the capacity (voltage rating) of a given-sized machine. This has been accomplished most often by lowering the operating temperature of the generator, using hydrogen, oil, or water as cooling mediums.

A main objective of the power engineer is to make as much use of available generated heat as possible by minimizing the amount of heat lost. The lower the heat loss, the greater is the system's thermal efficiency. From the formula

$$E = [(T_1 - T_2)/T_1] \times 100$$

[where E = thermal efficiency, in percent; T_1 = absolute temperature of entering steam into the engine; T_2 = absolute temperature of exhausted steam], the heat used by the system is measured by a temperature gradient. To maximize this gradient (and thus thermal efficiency), engineers have been striving for higher steam temperatures and pressures and for lower exhaust temperatures.

In practice, the simple steam generation system for producing electric energy described above is much more complex. Numerous refinements in the process have been made. Those occurring since 1950 are studied in chapter IV.

Appendix C

Description of Selected Electric Power
Generation Innovations, 1950-1970

The following exposition describes briefly the technological innovations in electric power generation from 1950 to 1970 listed in Table 4.1 that are not accounted for in chapter IV. The numbers accompanying each innovation refer to those in Table 4.1.

[4] *First pressurized boiler.* The use of pressurized boilers reduces the generating plant's capital cost by eliminating the need for induced-draft fans that pull the heated air out of the boiler into the stacks. The air pressure inside a pressurized boiler is greater than atomspheric pressure, allowing the boiler gases to pass out of the boiler into the stack naturally without force.

[5] *First welded-wall boiler.* Welded-wall boilers also reduce capital costs, as the employment of welded panels to surround the boiler tubes eliminates the need for steel casings to make a gas tight setting on pressurized boilers.

[6] *First combined-circulation boiler.* This design involves the recirculation of boiler tube fluid through the furnace wall system using a recirculation pump that automatically produces the required fluid velocities regardless of the unit's load. The fluid pressure drop between boiler sections is reduced in the combined-circulation boiler allowing lower operating pressures to be used. Greater fexlibility for variable pressure operation also is produced with this design.

[7] *First double reheat generating unit.* The reheating of steam in the boiler after it has passed through part of the turbine produces increased fuel cycle efficiency. After passing through the first several stages of the turbine, the steam is piped to a superheater in the boiler and reheated to approximately the initial steam temperature and then piped back to the remaining stages of the turbine. The reheat process increases the thermal range, and thus thermal efficiency, of the fuel cycle. With double reheat the steam is reheated

113

twice. Steam is removed twice from the turbine (each time at different stages) and twice reheated and piped back to the next stage, further improving the thermal efficiency of the operating cycle.

[8] *First high-speed boiler feed pump.* The boiler feed pump plays a critical role in steam-electric power generation. It circulates water from the steam-condensate circuit back to the boiler. As operating pressures and unit sizes have risen, more powerful and larger-capacity boiler feed pumps have become required. With the pumps' increased size more efficient methods of driving them had to be developed. The first high-speed boiler feed pump was driven by an 1,800 rpm constant-speed motor that drove the pump through step-up gearing at speeds over 8,300 rpm. The successful development of larger-capacity, higher-speed boiler feed pumps has allowed unit sizes and pressures to increase correspondingly.

[9] *First installation of main turbine boiler feed pump drive.* Another configuration of boiler feed pump drives is to power the pump by the main steam turbine-generator. Instead of using a separate motor to drive the boiler feed pump, the pump is connected to the turbine-generator shaft through an adjustable speed hydraulic coupling. The main turbine drive has been used by power engineers to reduce the amount of horsepower needed to drive the pumps.

[10] *First single boiler feed pump per unit design.* Since the boiler feed pump is a critical piece of equipment, most power stations use several pumps simultaneously so that the entire unit does not have to shut down if a pump malfunctions. Two one-half capacity or three one-third capacity pumps may be used with spare pumps on standby. As the boiler feed pumps and related equipment may cost over $2 million, the utilities have attempted to reduce this capital cost by reducing the required number of boiler feed pumps. The use of one full-capacity boiler feed pump eliminates the need for other operating pumps and reduces the plant's capital cost. Such a development was based on advances in pump design and operation that substantially increased the dependability and capacity of the boiler feed pumps. Another design change employed at the Glen Lynn Station was the use of an auxiliary steam turbine to drive the pump. The use of a single boiler feed pump driven by an auxiliary turbine increased the plant's capacity about 2 percent as less of the plant's generated electricity was used by the boiler feed pump equipment.

[11] *First brushless exciter.* An exciter is a separate generator that produces electric current needed to "excite" the field coils surrounding the main generator's rotor. The current in these coils magnetizes the electromagnet, which produces electricity as it rotates inside the stator windings. Without the exciter to produce the magnetic field in the rotor windings, the generator could not produce electricity. To provide the field windings with current, two collector rings (terminals) are mounted on the end of the main generator shaft. These rings are connected to either end of the wires com-

posing the field windings and rotate as the shaft turns. Stationary commutator brushes in contact with the collector rings conduct the exciting current from the exciter to the windings on the generator rotor. By eliminating the generator collector rings and commutator brushes, the principal cause of failure for the excitation system is removed, which increases the system's reliability. The brushless exciter contains no commutator brushes, no collector rings. Excitation system maintenance is also reduced. It produces current by means of a rotating rectifier coupled to an a.c. generator.

[13] *First generator with inner-cooling on both the rotor and stator.* This design is similar to the generator cooling system described in chapter IV [innovation (12)] except that now both the rotor and the stator are cooled. The hydrogen gas flows within the confines of the ground wall insulation surrounding both the rotor and stator. Such cooling methods have increased the ratings of a given-sized generator, lowering the unit capital costs.

[14] *First large liquid-cooled generator.* Instead of cooling the generator stator with hydrogen gas, the coolant used at Eastlake Station's generator was liquid mineral oil. The oil cooled the stator windings. Different coolants are used because of their varying heat-removal properties. The goal of liquid cooling, like other generator cooling methods, is to increase the generator's rating. This was accomplished at Eastlake Station, as the generator was originally rated at 260,000 kilovolt-amperes (kva). After liquid-cooling was used in the same-sized machine it was rerated at 305,000 kva.

[15] *First water-cooled generator in United States.* Engineers have known that water is much more effective in removing heat than hydrogen gas. The use of water as a generator stator coolant is more complicated than using hydrogen for three reasons: Because of (1) the necessity for installing a water treating and recirculation system; (2) the flexible insulating connecting tubes needed to introduce the coolant into the stator windings; and (3) the precautions that must be taken to prevent water leakage inside the stator windings. Its advantages are that it is nonflammable and universally available. The disadvantages center around its corrosive properties and the need for high purity. The technology was developed to overcome these obstacles, and water-cooling of generator stator windings was introduced in England at the Bold Power Station. Generator ratings increased as machine efficiency was improved through water-cooling.

[16] *First double-rotation turbine in United States.* The Detroit Edison Company was the first utility in the United States to use a double-rotation (radial-flow) steam turbine. Unlike other turbines where the steam flows axially along the turbine shaft, the steam in the radial-flow turbine flows outward from the center of the turbine shaft into the successive stages of turbine blades, arranged in concentric rows with increasing diameters. Detroit Edison purchased the Swedish-made turbine because of its compact design, and because it was the most economical unit available.

[17] *First large condensor using aluminum tubing.* Steam condensing units usually employ copper alloy tubing in the heat exchanger to cool and condense the steam. Other types of tubing have been tested. While aluminum alloy tubing was used experimentally and found to have no inherent technical advantages over copper alloy tubing, the material cost of aluminum was approximately one-third that of copper tubing, demonstrating a clear economic advantage.

[18] *First condensor completely tubed with stainless steel.* Stainless steel tubing was tested in areas where cooling water has a high mineral content to evaluate the resistance of the tubing to steam-side and water-side erosion and to determine any adverse corrosion or other conditions that might exist. Like aluminum tubing, the cost of stainless steel tubing was lower than that of copper.

[19] *First large side-entry condensor.* Unlike conventional condensors that are placed under the steam turbine, the side-entry condensor is connected to an axial exhaust flow turbine at the same level. The steam exits the turbine and enters the condensor horizontally instead of vertically. Advantages to this arrangement include simplified piping and better utilization of space.

[21] *First canned motor pump for boiler feed water circulation.* The canned motor pump was originally developed for use in atomic submarine power plants. The advantages of such a design center on the pump's greater reliability and more compact design. The pump operates within the feed water recirculation system, not externally as in conventional designs. Because of the internal design, the amount of required piping needed in the system is reduced, representing a savings in capital cost.

[22] *First use of a bag filterhouse to reduce stack emissions.* The bag filterhouse system is composed of two principal pieces of equipment: (1) a regenerative-type air preheater and (2) a two-compartment baghouse. The bag (about 28 feet long) is made of siliconized fiberglass cloth. The baghouse uses continuous additions of solid alkaline additives to remove virtually all of the SO_3 from the stack gases. The gases pass through the preheater after leaving the boiler and then enter the baghouse where the visible stack emissions and sulfur oxide are filtered out.

[23] *First use of two-stage combustion system.* Control of the nitrogen oxide emissions in stack gases from oil-fired plants can be accomplished by using a two-stage combustion system. The result of this combustion system is lowered combustion temperatures which reduce the formation of nitrogen oxides in the flue gases.

[24] *First pilot plant use of Wellman-Lord SO_2 removal process.* This process utilizes a wet scrubbing system combined with a chemical reaction plus a change in chemical form to remove the sulfur dioxide in the stack gases. It is a regenerative process for the recovery of pure anhydrous liquid

SO_2 and elemental sulfur or sulfuric acid. According to the manufacturer, Wellman-Lord, Inc., it is the first chemical SO_2 recovery process to give a favorable economic return.

[28] *First central station to use automatic data logging.* By employing automated controls to record performance data, the plant operators have more complete and detailed current knowledge of plant operations. The operators can more easily adjust plant procedures to increase plant performance. This automatic control installation cost Gulf States Utilities $150,000. The installed logging system reduced the required number of plant operators (they were reassigned to run the utility's new generating plants). The installation was expected to pay for itself in three years.

[29] *First U.S. power plant with data logging, scanning, and alarming function control system.* This innovation in power plant controls represents the next stage toward fully automated control after (28). The scanning and alarming functions automatically provide the plant operators with current information about the status of the plant's operating systems and warn the operators when a system malfunction occurs.

[33] *First automation of a 2,400 psi, reheat, once-through generating unit.* The more sophisticated the steam cycle becomes, the more complex is the plant control system. This innovation demonstrates the rapid development of automatic control equipment technology. Southern California Edison successfully automated an oil-fired generating unit having a modern steam cycle design.

[34] *First TV viewing of boiler combustion.* Predating the fully automated combustion control systems employing computer technology, the use of television to allow plant operators to watch the combustion process in the boiler greatly improved operating efficiency and dependability.

[35] *First regenerative gas turbine in central station use.* The regenerative-cycle gas turbine, unlike a simple-cycle gas turbine, uses part of the heat generated in the exhaust gases to preheat the air entering the combustion chamber. By employing this heat again in the cycle the turbine's efficiency is raised.

[36] *First automatic "self-starter" for a gas turbine in central station use.* The development of an automatic self-starter for gas turbines enabled the utilities to operate the gas turbines by remote control. This gave the utilities more flexibility in system design, operation, and siting.

[39] *First coal gas-fired gas turbine to operate in a combined cycle with a steam turbine.* Unlike the combined cycle designs discussed in chapter IV, this cycle uses coal gas as the fuel for the gas turbine. Like other gas turbine-steam turbine cycles, the design creates greater thermal efficiencies by reducing the amount of nonutilized heat.

[40] *First jet gas turbine repowering of generation units.* A low-cost peaking and emergency power station was installed by replacing old conventional steam turbines and boilers with jet gas turbines. The previously used generators and other station equipment continued to be utilized. The generators were repowered with the gas turbines to provide Louisville Gas and Electric with standby power. Since the existing plant structure and generators were not replaced, this method of creating new generating facilities substantially reduced construction and capital costs.

[41] *First pipeline transportation of coal.* The first conveying of coal by other than wheeled vehicles to a power station was developed to reduce the transportation cost of coal. Cleveland Electric Illuminating Company and Pittsburgh Consolidated Coal Company realized that in order to retain their competitive position, cheaper means of coal transport must be developed. Railroad freight rates were then almost equal to the cost of the fuel at the mine. The coal slurry pipeline was built at a cost of $25 million. It was operated, maintained, and owned by the Pittsburgh Consolidated Coal Company. The pipeline went into operation in 1957. The transportation cost of the coal using the pipeline was between $2.47 and $2.65 per ton. It was shut down in 1963 when the railroads, realizing that they were losing business to the pipeline, offered a rate of $1.88 per ton using a shuttle train. The railroads acknowledged that the coal pipeline spurred thinking in their industry as to how the railroads could again get Cleveland Electric Illuminating's business. (The pipeline was used in the reverse direction to carry a mixture of 95 percent water and 5 percent solids of dredge material, fly ash, and sludge with which to fill up barren areas left by strip mining when it was no longer used to carry coal to Cleveland.)

[42] *First use of integral train concept of shipping coal to a power station.* By using semipermanently coupled integral trains to carry 6,000 tons of coal in seventy cars over a 400-mile route, the railroads found an answer to the competitive threats posed by coal pipelines and mine-mouth generating stations. The economics realized from such integral trains allowed the railroads to reduce rates by as much as $1.04 per ton.

[Innovations 43 through 47 are self-explanatory. Technological developments in metallurgy and high-temperature physics have allowed operating temperatures and pressures to increase significantly over the past two decades.]

[49] *First commercial-sized nuclear power plant.* On 22 March 1955 Consolidated Edison filed the first license application for a central station nuclear power plant of commercial size to be built without government funds. The Indian Point Atomic Power Plant, a 163 Mwe PWR, finally began commercial power generation on 16 September 1962. The Indiana Point Plant was the first commercial thorium-uranium converter to go into operation in the United States. Total costs for the plant, including R&D, were $134 million, over 2.5 times the original cost estimate. The total capacity of the station is 275 Mwe with two oil-fired superheaters generating 112 Mwe in ad-

dition to the reactor's output. The management of Consolidated Edison has stated that, if the power industry is to get additional experience with nuclear power, a utility ought to build an atomic power plant as soon as nuclear technology makes it economically comparable to the fossil-fueled plants on the utility's system. The total energy cost for the Indian Point Plant was 14.7 mills per kwh.

[51] *First nuclear power plant with nuclear superheat.* One of the problems of nuclear reactor designs was the poor steam conditions produced by the reactor. The low steam temperatures and pressures severely reduced the thermal efficiency of the reactor. Great efforts were made to develop new reactor designs that would produce better quality steam and thus raise operating efficiencies. The Pathfinder Nuclear Power Plant was the first light water reactor (a BWR) to incorporate an integral nuclear superheater that would raise the steam temperature from 336°F to 825°F before the steam was piped to the turbine. At the time, great interest was given to nuclear superheat as a way of increasing efficiencies. The Pathfinder Plant went critical in March 1964. It was idled in September 1967 from damage to its steam separator nozzles. After several attempts to improve the components' design, the reactor was shut down in November 1967. During the period from 1965 to 1967 there was decreasing interest in nuclear superheat reactors resulting from the progress in light water reactor design that increased operating efficiencies without superheat.

[52] *First organic-cooled and -moderated reactor.* Organic-cooled and -moderated reactors offered another potential design for competitive nuclear power. The simplicity of the reactor design and the direct applicability of technology for water reactor designs made the organic-cooled and -moderated reactor concept desirable. These reactors operate at higher temperatures and lower pressures than light water reactors. They have no corrosion problems, making for low capital cost. The coolant and moderator, usually a hydrocarbon like Terphenyl, is expensive to replace when it has decomposed and has poor heat transfer characteristics. The Piqua reactor was built by the AEC for the city of Piqua, Ohio, under the Power Demonstration Reactor Program. It was a small, 11.4 Mw, experimental reactor. The estimated cost of the project was $16 million, including construction and R&D. The reactor started operation in August 1962. By 1963 the AEC had decided to give up on the organic-cooled and -moderated reactor concept as it had little potential for producing competitive power. The Piqua plant was closed down in 1967.

[53] *First sodium-graphite-moderated thermal reactor.* The sodium-graphite reactor design is complex and expensive to build but operates at high temperatures and thermal efficiencies. To be economical the fuel elements of such a reactor design have to be capable of long life at high temperatures. The 75 Mw Hallam Power Facility, owned by the AEC, provided design, construction, and operating experience with a prototype sodium-cooled reactor. The reactor was producing power in 1963; however, in September 1964

the facility was shut down when seven of the 141 stainless clad graphite moderator blocks failed. In August 1965 the AEC and Consumers Public Power District terminated their contract to operate the reactor.

[54] *First heavy water-cooled and -moderated pressure tube reactor.* The heavy water-cooled and -moderated reactor design benefits from using natural uranium as the fuel, and has no obvious size limitations. The disadvantages lie in using costly heavy water as the coolant and moderator. The Parr Shoals project, designed to demonstrate the feasibility and economic potential of the concept, was built and operated by the Carolinas Virginia Nuclear Power Associates. The estimated cost was $28 million, $14 million of which was paid by the AEC. The 19 Mwe reactor began operation in May 1964. Because of the advances in other reactor designs and the Parr Shoals project's high capital costs, further consideration of the heavy water reactor concept was not undertaken. The Parr Shoals plant was shut down in January 1967.

[56] *First jet-type water pumps within the pressure vessel.* The use of jet pumps within the reactor's pressure vessel provides water recirculation with a minimum of external loops (piping outside the pressure vessel). These pumps reduce the requirement for primary water loops (circuits), thus reducing the amount of piping needed in the reactor. They also are more compact, more reliable, and provide more safety than pumps located outside the pressure vessel.

[57] *First use of noncanned motor pump unit in a large PWR.* After studying the advantages of using canned pumps in a PWR, which center around their hermetically sealed construction, the designers of the San Onofre Plant (Westinghouse and Betchel Corporation) decided to use a conventional, noncanned pump. By employing a standard a.c. induction motor instead of a canned motor, the system's electrical efficiency was increased by approximately 8 percent. Thus, despite the more complicated design, the conventional pump was more efficient from an engineering and economic standpoint and thus preferred.

[58] *First commercial reactor using a prestressed concrete pressure vessel.* The advantages of using prestressed concrete in the pressure vessel are largely economic. The prestressed concrete pressure vessel easily can be constructed at the reactor site, avoiding the size and transportation difficulties of large steel vessels. The on-site construction produces large cost savings. There are also safety advantages in using prestressed concrete. The tension cables located in the outer regions of the concrete walls assure structural integrity and resistance to cracking. Such vessels are made leakproof by a steel liner, which serves as the inner form for pouring the concrete. Public Service of Colorado's Fort Saint Vrain Reactor will use a prestressed concrete pressure vessel and realize these cost and safety benefits.

[59] *First reinforced concrete pressure containment structure.* The advantages of using a concrete containment structure are similar to those discussed above in innovation (58). The on-site construction of a concrete containment structure is simpler and less costly than a similar-sized steel containment structure.

Appendix D

Selected Technological Innovations in
Electric Power Generation
Before 1950

Boilers, Boiler Design
(1) First use of pulverized coal in a power plant, Oneida Station, Milwaukee Electric Railway & Light Company, 1918.
(2) First cyclone furnace, Calumet Station, Commonwealth Edison Company, 1944.
(3) First high-pressure boiler in United States utilizing controlled circulation (2,000 psig, 960°F), Somerset Station, Montaup (Massachusetts) Electric Company, 1942.
(4) First high-pressure, natural-circulation generating unit (2,300 psi), Twin Branch 3, Indiana & Michigan Electric Company, 1941.
(5) First large boiler-turbine-generator of unit design, Port Washington Station, Wisconsin Electric Power Company, 1935.
(6) First reheat generating unit, Philo 1, Ohio Power Company, 1924.

Boiler Feed Pumps (BFP), Turbines, Generators
(7) Operation of first steam power units with 3,600 rpm motor and hydraulic drives for BFP's, Essex Station, Public Service Electric and Gas Company, 1947.
(8) First commercial hydrogen-cooled generator, Frank M. Tait Station, Dayton Power and Light Company, 1937.

Condensors, Cooling Towers, Water Purification
(9) First evaporator in distillation of make-up (water), Connors Creek Station, The Detroit Edison Company, 1915.
(10) First installation of make-up evaporator between two successive extraction points of the turbine, Saginaw River Station, Consumers Power Company, 1923.
(11) First use of carbonaceous zeolites (ion-exchange methods of demineralization) Gulf States Utilities Company, 1937.
(12) First electric plant using demineralization (versus evaporation), Muscatine Municipal Power Plant, Muscatine, Iowa, 1948.

Pollution Abatement Equipment

(13) First use of electrostatic precipitators, Trenton Channel Station, The Detroit Edison Company, 1924.

Automatic Control Equipment

(14) Pioneering use of a centralized combustion control system, Sherman Creek Station, Consolidated Edison Company, 1922.

(15) First use of centralized station control, Oswego Steam Station, Niagara Mohawk Power Corporation, 1942.

(16) First use of a network analyzer by an electric utility, Public Service Electric and Gas Company, 1938.

Gas Turbines, Combined Cycle Units

(17) First stationary, central station, gas turbine generating unit (operating on natural gas), Arthur S. Huey Generating Station, Oklahoma Gas and Electric Company, 1949.

(18) First mercury-steam binary cycle (topping) unit, Dutch Point Station, Hartford (Connecticut) Electric Light Company, 1922. (First commercial full-scale, mercury-steam unit, South Meadow Station, Hartford Electric Light Company, 1928.)

Operating Temperatures and Pressures

(19) First commercial generating station to operate at 1,000°F (at 1,350 psi), Missouri Avenue Plant, Atlantic City Electric Company (then part of AEP), 1946.

Appendix E

Future Power Generation Methods

Magnetohydrodynamics (MHD)

Power generation by MHD is based on Michael Faraday's discovery in 1931 of electromagnetic induction—when a conductor and a magnetic field move with respect to each other, an electric voltage is induced in the conductor. The MHD generator extracts mechanical energy from the moving conducting field and delivers an electrical energy output. The flow of electricity moves across the magnetic field as a stream of ionized gas passes through the field. The cycle compresses atmospheric air to about 140 psia and heats it to 3,600°F in a regenerative air heater. Coal combustion raises the gas to 5,300°F. This high-temperature gas (its conductivity is increased when necessary by "seeding," that is, by adding a small amount of vaporized metal such as potassium) then passes through the MHD generator developing about 360 Mw d.c. energy. The nuclear MHD cycle is similar except that a helium stream seeded with cesium is compressed, then heated in a regenerator to 2,500°F. The nuclear reactor adds the next 1,600 degrees and discharges the ionized helium into the MHD generator at 4,100°F. Projected thermal efficiences for an MHD generator range from 55 percent (with a fossil-fueled heat source) to over 60 percent (with a nuclear heat source). In this cycle the steam turbines, boiler, compressor, a.c. generator, feedwater heaters and their auxiliaries are essentially conventional equipment and represent no problem in development. The MHD generator is essentially a rectangular, diverging nozzle some sixty feet long, expanding from three feet in diameter at the throat to six feet at the outlet. The electrodes, exposed to the supersonic plasma, are the generator's most critical components. The cathode must be a good electron emitter, and the anode must be a reasonably good conductor of electricity at high operating temperatures. This MHD equipment presents special technical problems that have yet to be fully solved. They require further R&D efforts, particularly on the behavior of high-temperature gases in magnetic fields and high-temperature resistant materials. In August 1959 an MHD generator constructed at the Avco-Everett Laboratory produced more than 10 kw of electric power.[1]

Thermionic Generation

Using the process of thermionic emission, electric energy is produced directly from heat energy without the use of moving mechanical equipment. Thermionic emission is the phenomenon of electron emission from the surface of an electron-conducting material due to the thermal energy of the electrons within the material. This phenomenon was first observed by Thomas Edison (and thus named the Edison Effect) in 1883, when electron emission was discovered from metals heated to high temperatures. Developments in the direct conversion of heat to electricity by this means have followed two research efforts at Los Alamos Laboratory: (1) design of a cesium-filled thermionic converter that operated in a reactor and produced about 40 watts of power; (2) similar work carried out with high vacuum, close-spaced diodes producing outputs of several watts using various heat sources, one being the concentrated rays of the sun.[2]

Thermoelectric Generation

As in thermionic generation, thermoelectric generation produces electric energy directly from thermal energy. The basic discovery or invention of thermoelectricity was made by accident in 1820 by Thomas J. Seebeck. The Seebeck Effect states that if a closed circuit comprises two dissimilar metals, and if a temperature differential is maintained between the two, a current will flow. The thermocouple is founded on this principle. As steam engines improved in efficiency, the thermocouple was relegated more and more to the background until it became nothing more significant than a temperature-measuring device. A number of theoretical developments in solid-state physics and quantum mechanics and the development of devices like the semiconductor have now made it possible to think seriously of the thermocouple as a practical device for converting heat to electricity.[3]

Fuel Cell

The fuel cell is a device for converting chemical energy directly into electric energy. The concept of the fuel cell and demonstration of its principle have been known since the mid-1880s. The fuel cell resembles a conventional storage battery. Unlike the storage battery, which needs to be recharged, the fuel cell operates as long as fuel and oxygen are fed into the electrodes. The oxygen ions (coming from air at the air electrode) migrate through the electrolyte, combine with the fuel (hydrogen and carbon monoxide) to form water and carbon dioxide, and deposit the two electrons at the fuel electrode, giving it a negative charge. When the external circuit is closed, the electrons flow from the fuel electrode to the air electrode. Fuel cells are series-connected to form a battery. Batteries are then series-connected to form a d.c. power package of 800 to 2,000 volts. Fuel cells have been developed to operate with the products of partially oxidized coal, oil, and gas. The mixtures of carbon monoxide and hydrogen can be pretreated before going into the cells to remove particulates and the compounds of sulfur. The heat generated in the fuel cells could be recovered in the gasification cycle of coal, eliminating the need for cooling and consequent thermal pollution. Projected thermal efficiencies for such plants are from 60 to 70 percent, including fuel preparation.[4]

Appendix F

Privately Owned Electric Utilities'
Total Factor Productivity
1953–1970

The following methodology was used to determine the yearly total factor productivity from 1953 to 1970 for electric utilities. These measurements update John Kendrick's findings about electric utilities' total factor productivity, which ended with 1953 data. Data were available only for the private sector of the electric utility industry, but should be representative of the entire industry. The source for these data was the *EEI Pocketbook of Electric Utility Industry Statistics*, 17th ed. (1971).

Using the formula for total factor productivity

$$P = \frac{Q}{eL + fC}$$

[where P = total factor productivity; Q = output, as a percentage of output in some base period; L = labor input, as a percentage of labor input in some base period; e = labor's share of the value of the output in some base period; C = capital input, as a percentage of capital input in some base period; and f = capital's share of the value of the output in some base period], the total factor productivity index for privately owned electric utilities was determined. The base year selected was 1953. The labor and capital shares of the value of output are taken from Kendrick, and are the relative average values for unit compensation of each factor. In 1953 labor's share was 58 and capital's share was 42.

The figures for total factor productivity presented in Table F-2 should be taken as rough estimates; no attempt was made to update the factor shares from the 1953 data of Kendrick. It would not be unexpected to find that the capital share has increased relative to labor's share since 1953.

Table F-1 presents the privately owned electric utilities' output (in billions of kwh), capital (as measured by utility plant investment in millions of dollars), and labor (the average number of employees in operation, maintenance, and construction) for each of the years since 1953. The yearly indexes for total factor productivity, presented in Table F-2, are derived by substituting the figures given in Table F-1 into the total factor productivity formula with 1953 as the base year.

Table F.1. *Privately Owned Electric Utilities' Output, Capital, and Labor Statistics, 1953–1970*

Year	Output	Capital	Labor
1953	354.3	28,935	336,900
1954	371.0	31,562	339,000
1955	420.9	34,049	339,300
1956	459.0	36,814	344,400
1957	480.9	40,395	350,000
1958	490.4	44,255	348,300
1959	544.2	47,134	344,300
1960	578.6	50,592	345,900
1961	604.9	53,731	343,000
1962	651.0	56,756	339,800
1963	701.3	59,861	339,500
1964	756.2	63,040	341,500
1965	809.5	66,863	345,400
1966	880.8	71,664	349,100
1967	928.4	77,607	357,400
1968	1,019.3	84,468	365,300
1969	1,102.2	92,466	372,500
1970	1,182.9	102,447	384,900

SOURCE: *EEI Pocketbook of Electric Utility Industry Statistics,* 17th ed., 1971, pp. 6, 14, and 39.

NOTE: Output measured in billions of kwh. Capital represented by utility plant investment in millions of dollars. Labor represented by the average number of employees in operation, maintenance, and construction.

Table F.2. *Privately Owned Electric Utilities' Total Factor Productivity Indexes 1953–1970*

Year	Productivity index	Year	Index
1953	1.00	1962	1.30
1954	1.00	1963	1.36
1955	1.09	1964	1.42
1956	1.16	1965	1.46
1957	1.14	1966	1.52
1958	1.11	1967	1.51
1959	1.21	1968	1.55
1960	1.22	1969	1.57
1961	1.25	1970	1.55

The figures presented in Table F-2 indicate that total factor productivity for the privately owned electric utilities has risen by 55 percent since 1953. This represents an average (compound) yearly growth rate of 2.6 percent.

Appendix G

Selected Nuclear Utility Groups

*Yankee Atomic Electric Co. (Yankee)**

Boston Edison Co.
Cambridge Electric Co.
Central Maine Power Co.
Central Vermont Public Service Corp.
The Connecticut Light and Power Co.
The Hartford Electric Light Co.

Montaup Electric Co.
New Bedford Gas and Edison Light Co.
New England Power Co.
Public Service Co. of New Hampshire
Western Massachusetts Electric Co.

Atomic Power Development Associates (Enrico Fermi)

Alabama Power Co.
Atlantic City Electric Co.
Baltimore Gas and Electric Co.
Boston Edison Co.
Central Hudson Gas & Electric Corp.
Cincinnati Gas & Electric Co.
Cleveland Electric Illuminating Co.
Connecticut Light and Power Co.
Consolidated Edison Co. of New York.
Consumers Power Co.
Detroit Edison Co.†
General Public Utilities Corp.
Georgia Power Co.
Gulf Power Co.
The Hartford Electric Light Co.
Long Island Lighting Co.
Metropolitan Edison Co.

Mississippi Power Co.
New England Power Co.
New Jersey Power & Light Co.
New York State Electric & Gas Corp.
Niagara Mohawk Power Corp.
Pennsylvania Electric Co.
Philadelphia Electric Co.
Potomac Electric Power Co.
Public Service Electric and Gas Co.
Rochester Gas and Electric Corp.
Southern Services, Inc.
Toledo Edison Co.
Wisconsin Electric Power Co.
Wisconsin Power and Light Co.
Allis-Chalmers Mfg. Co.
Babcock & Wilcox Co.
Bendex Aviation Corp.
Commonwealth Associates, Inc.

129

Ford Motor Co.
General Motors Corp.
Jackson & Morland, Inc.
Negea Service Corp.

Pittsburgh Consolidation Coal Co.
United Engineers & Constructors, Inc.
Vitro Corporation of America

Carolinas Virginia Nuclear Power Associates (Parr Shoals)

Carolina Power & Light Co.
Duke Power Co.

South Carolina Electric & Gas Co.†
Virginia Electric and Power Co.

Nuclear Power Group (Dresden)

American Electric Power Service Corp.
Central Illinois Light Co.
Commonwealth Edison Co.†

Illinois Power Co.
Kansas City Power & Light Co.
Pacific Gas and Electric Co.
Union Electric Co.

High Temperature Reactor Development Associates (Peach Bottom)

Alabama Power Co.
Arizona Public Service Co.
Arkansas Power & Light Co.
Atlantic City Electric Co.
Baltimore Gas and Electric Corp.
California Electric Power Co.
Central Illinois Electric and Gas Co.
Central Illinois Light Co.
Central Illinois Public Service Co.
Central Louisiana Electric Co., Inc.
Central Power and Light Co.
Cincinatti Gas & Electric Co.
Cleveland Electric Illuminating Co.
Delmarva Power & Light Co.
The Detroit Edison Co.
Gulf Power Co.
Gulf States Utilities Co.
The Hawaiian Electric Co., Ltd.
Idaho Power Co.
Illinois Power Co.
Iowa Public Service Co.
Kansas City Power & Light Co.
Kansas Power and Light Co.
Kentucky Utilities Co.
Louisiana Power & Light Co.
Mississippi Power Co.
Mississippi Power & Light Co.

Missouri Public Service Co.
The Montana Power Co.
New Orleans Public Service Inc.
New York State Electric & Gas Corp.
Niagara Mohawk Power Corp.
Pacific Gas and Electric Co.
Pacific Power & Light Co.
Pennsylvania Power & Light Co.
Philadelphia Electric Co.†
Portland General Electric Co.
Public Service Co. of Colorado
Public Service Co. of New Mexico
Public Service Co. of Oklahoma
Public Service Electric and Gas Co.
Puget Sound Power & Light Co.
Rochester Gas and Electric Co.
St. Joseph Light & Power Co.
San Diego Gas & Electric Co.
Sierra Pacific Power Co.
Southern California Edison Co.
Southwestern Electric Power Co.
United Illuminating Co.
Utah Power & Light Co.
Washington Water Power Co.
West Texas Utilities Co.

Central Utilities Atomic Power Associates (Pathfinder)

Central Electric & Gas Co.
Interstate Power Co.
Iowa Power and Light Co.
Iowa Southern Utilities Co.
Madison Gas and Electric Co.
Mississippi Valley Public Service Co.

Northern States Power Co.†
Northwestern Public Service Co.
Otter Tail Power Co.
St. Joseph Light & Power Co.
Wisconsin Public Service Corp.

---

*Name of the nuclear power plant developed and constructed by the group.
†Indicates the utility that heads the group.

Appendix H

Selected Major Domestic Power
Equipment Manufacturers

Company	*Major items of equipment supplied to the electric utility industry*
General Electric Co.	Turbine-generators, motors, transformers, pumps, reactors, meters, rectifiers, and controls
Westinghouse Electric Co.	Turbine-generators, motors, transformers, pumps, reactors, meters, rectifiers, controls, heat exchangers, and blowers
Combustion Engineering, Inc.	Boilers and reactors
Babcock & Wilcox Co.	Boilers and reactors
Allis-Chalmers Mfg. Co.	Transformers, pumps, motors, deaerators, and reactors
General Dynamics Corp.; General Atomic Division	Reactors
North American Aviation, Inc.; Atomics International Div.	Reactors
Foster Wheeler Corp.	Boilers, heat exchangers, and pumps
Worthington Corp.	Heat exchangers, pumps, condensors, and compressors
Anaconda Wire & Cable Co.	Cable and cable accessories
The Okonite Co.	Cable and cable accessories
General Cable Corp.	Cable and cable accessories
McGraw-Edison Co.	Transformers and fuses
I-T-E Circuit Breaker Co.	Switch equipment and bus duct

SOURCE: *EEI Bulletin* 32 (July 1964) :166.

Appendix I

The Fifty Largest Privately Owned Electric Utility Systems and Affiliated Operating Utilities

(1) The Southern Co.: Alabama Power Co., Southern Electric Generating Co., Gulf Power Co., Georgia Power Co., Mississippi Power Co.

(2) American Electric Power Co.: Indiana & Michigan Electric Co., Kentucky Power Co., Ohio Power Co., Kingsport Power Co., Appalachian Power Co., Wheeling Power Co., Michigan Power Co.

(3) Commonwealth Edison Co.: Commonwealth Edison Co. of Indiana, Commonwealth Edison Co.

(4) Southern California Edison Co.

(5) Pacific Gas and Electric Co.

(6) Duke Power Co.

(7) Texas Utilities Co.: Dallas Power & Light Co., Texas Electric Service Co., Texas Power & Light Co.

(8) Middle South Utilities Co.: Arkansas Power & Light Co., Louisiana Power & Light Co., New Orleans Public Service Co., Mississippi Power & Light Co.

(9) Detroit Edison Co.

(10) Consolidated Edison Co.

(11) Public Service Electric and Gas Co.

(12) Houston Lighting & Power Co.

(13) Central & South West Corp.: Public Service Co. of Oklahoma, Central Power and Light Co., Southwestern Electric Power Co., West Texas Utilities Co.

(14) Florida Power & Light Co.

(15) Virginia Electric and Power Co.

(16) General Public Utilities Corp.: Jersey Central Power & Light Co., New Jersey Power & Light Co., Metropolitan Edison Co., Pennsylvania Electric Co.

(17) Allegheny Power System: West Penn Power Co., Potomac Edison Co.

133

of Pa., Potomac Edison Co., Potomac Edison Co. of W. Va., Potomac Edison Co. of Va., Monongahela Power Co.

(18) Philadelphia Electric Co.: Susquehanna Electric Co., Conowingo Power Co., Philadelphia Electric Power Co., Philadelphia Electric Co.
(19) Consumers Power Co.
(20) Gulf States Utilities Co.
(21) Carolina Power & Light Co.
(22) Ohio Edison Co.: Pennsylvania Power Co., Ohio Edison Co.
(23) Niagara Mohawk Power Corp.
(24) Union Electric Co.: Missouri Power & Light Co., Union Electric Co.
(25) Ohio Valley Electric Corp.: Indiana-Kentucky Electric Corp., Ohio Valley Electric Corp.
(26) Pennsylvania Power & Light Co.
(27) Potomac Electric Power Co.
(28) Baltimore Gas and Electric Co.
(29) Wisconsin Electric Power Co.: Wisconsin–Michigan Power Co., Wisconsin Electric Power Co.
(30) Northeast Utilities: Holyoke Water Power Co., Holyoke Power & Electric Co., Western Massachusetts Electric Co., Connecticut Light and Power Co., Hartford Electric Light Co.
(31) Northern States Power Co.: Northern States Power Co. (Minnesota), Northern States Power Co. (Wisconsin).
(32) Cleveland Electric Illuminating Co.
(33) New England Electric System: New England Power Co., Massachusetts Electric Co., Granite State Electric Co., Narragansett Electric Co.
(34) Public Service Co. of Indiana
(35) Long Island Lighting Co.
(36) Oklahoma Gas and Electric Co.
(37) Duquesne Light Co.
(38) Florida Power Corp.
(39) Illinois Power Co.
(40) Boston Edison Co.
(41) Cincinatti Gas & Electric Co.: Union Heat, Light and Power Co., Cincinatti Gas & Electric Co.
(42) Idaho Power Co.
(43) South Carolina Electric & Gas Co.
(44) Indianapolis Power & Light Co.
(45) Public Service Co. of Colorado: Cheyenne Light, Fuel and Power Co., Public Service Co. of Colorado.
(46) Southwestern Public Service Co.
(47) New York State Electric & Gas Corp.
(48) Northern Indiana Public Service Co.
(49) Pacific Power & Light Co.
(50) Arizona Public Service Co.

NOTE: Utilities ranked by 1970 kwh generated.

Notes

CHAPTER I

1. National Science Foundation, *Employment of Scientists and Engineers in the U.S., 1950-1970* (Washington, D.C.: Government Printing Office, 1973).

2. Abbot P. Usher, *A History of Mechanical Invention* (New York: McGraw-Hill, 1929).

3. M. Abramovitz, "Resources and Output in the United States since 1870," *American Economic Review* 46 (May 1956): 5-23; R. M. Solow, "Technical Change and the Aggregate Production Function," *Review of Economics and Statistics* 39 (August 1957): 312-20.

4. Edwin Mansfield, *Industrial Research and Technological Innovation* (New York: W. W. Norton & Co., 1968); *The Economics of Technological Change* (New York: W. W. Norton & Co., 1968); and "Innovation and Technical Change in the Railroad Industry," in *Transportation Economics*, a Conference of the Universities—National Bureau Committee for Economic Research (1965), pp. 169-97.

5. The following list of pioneering, large-scale electric generating units illustrates the point that only the largest, interconnected utility systems have the option of installing new-developed, large-scale units.

Pioneering, Large-Scale Generating Units

Date of operation*	Unit size (Mw)	Plant name	Utility
		Kanawha	Appalachian
1/53	217	River	Power Co. (AEP)
11/55	250	Gallatin	TVA
10/57	300	River Rouge	Detroit Edison Co.
			Indiana & Michigan
7/59	450	Breed	Electric Co. (AEP)
		Widows	
9/60	500	Creek	TVA
9/62	650	Paradise	TVA
			Consolidated Edison
6/65	1000	Ravenswood	Co.

* As reported in *Power* Magazine's annual Steam Plant Design Survey.

6. *Invention* has been defined in various ways. Some analysts have defined an invention as an increase in knowledge embodying a new product or process for producing goods. Others have defined the invention according to its initial effect on relative factor shares. Still other researchers have added that the product or process must have prospective utility in addition to novelty if it is to be an invention. Whichever definition is used, invention precedes innovation in the process of technological advance. *Innovation* is defined in this study to be the first installation and commercially productive operation of a new process or piece of equipment.

7. The following trade journals and engineering society papers have been reviewed to collect a list of technological innovations in the areas of fossil-fueled and nuclear generation of electric power between 1950 and 1970: *Electrical World, Nucleonics, Power, Power Engineering,* and *Combustion,* as well as proceedings and papers of the American Power Conference, the American Society of Mechanical Engineers (ASME), and the Institute of Electrical and Electronic Engineers (IEEE).

8. The members of the panel who examined the list of innovations for accuracy, completeness, and technological significance are: Robert B. Boyd, Bureau of Power, Federal Power Commission; J. J. O'Connor, editor-in-chief, *Power* magazine; J. Guy Farthing, senior staff editor, *Electrical World;* Professor A. H. El-Albiad, Department of Electrical Engineering, Purdue University; and S. W. Shields, Richard Willis, and John Bott of the Engineering Department of Public Service Company of Indiana.

CHAPTER II

1. Changes in the retail price of electricity are compared to changes in the consumer price index in the following table. (1967=100)

	1950	1960	1970
Retail price index of electricity	90.8	99.8	106.2
Consumer price index, all items	72.1	88.7	116.3

SOURCE: U.S. Bureau of the Census, *Statistical Abstract of the United States, 1971,* 92d ed. (Washington, D.C.: Government Printing Office, 1971), tables 534 and 543, pp. 339 and 345.

2. Federal Power Commission, *National Power Survey, 1964,* pt I, p. 11.

3. Federal Power Commission, *Statistics of Privately Owned Electric Utilities in the United States, 1970,* p. ix. Hereafter referred to as *Statistics, Privately Owned.*

4. These figures represent only employees working for privately owned utilities. *Edison Electric Institute Pocketbook of Electric Utility Industry Statistics,* 17th ed. (1971), p. 39. Hereafter referred to as *EEI Pocketbook.*

5. Ibid., p. 13.

6. FPC, *Statistics, Privately Owned, 1970,* p. xii.

7. FPC, *National Power Survey, 1964,* pt I, p. 11.

8. FPC, *Statistics, Privately Owned, 1970,* p. xiii.

9. FPC, *National Power Survey, 1964,* pt I, p. 10.

10. *EEI Pocketbook,* pp. 17 and 18.

11. U. S. Department of Commerce, Bureau of Economic Analysis, *Survey of Current Business* 30 (December 1950): S-1, and 50 (December 1970): 8.

12. John Kendrick, *Productivity Trends in the United States* (Princeton: Princeton University Press, 1961), table 54, p. 137.

13. See Appendix F for methodology and computation used to calculate this figure and other total factor productivity data.

14. In 1950 and 1960 Class A utilities, according to the FPC, were those with annual electric operating revenues of $750,000 or more, and Class B utilities were those with annual electric operating revenues of $250,000 or more, but less than $750,000. By 1970 these classifications had changed; utilities with annual electric operating revenues of $2.5 million or more were in Class A, and those with annual electric operating revenues between $1 million and $2.5 million were in Class B.

15. FPC, *Statistics, Privately Owned, 1968,* p. v.

16. FPC, *National Power Survey, 1970,* p. I-2-5.

17. These 1,369 systems only distribute power; they do not generate or transmit any power for their service market. There are three facets to the electric utility industry: generation, transmission, and distribution. The generation of electric power, as its name implies, involves the production of electric energy. The process of steam electric power generation is described in Appendix B. The transmission of energy occurs at high voltages (some newer transmission lines carry current at 750,000 volts) from the generation source to a local distribution point, where the voltage is reduced for local distribution. Electric energy is distributed at the lower voltages from the transmission substation to the ultimate customer.

18. This figure is about one-tenth of the load density in urban areas. FPC, *National Power Survey, 1970,* p. I-2-6.

19. Ibid.

20. Ibid., p. I-2-4.

21. These figures represent the total number of operation, maintenance, and construction employees. *EEI Pocketbook,* p. 39.

22. Ibid., p. 13. Investment is measured by the amount of gross electric utility plant expenditures. Arthur D. Little, Inc. estimates the electric utilities' figures for net plant and equipment per employee to be $190,000, the highest of any industry examined. Bruce C. Old, *Trends in Research and Development* (Boston: Arthur D. Little, Inc., 1972), p. 26.

23. *EEI Pocketbook,* p. 15.

24. FPC, *Steam-Electric Plant Construction Cost and Annual Production Expenses,* Twenty-Second Annual Supplement—1969, p. 1-150, *passim.* Hereafter referred to as *Steam Plant.*

25. For the privately owned utilities the cost of gas and oil consumed is transformed into coal equivalent terms, usually by measuring the cost per BTU per cubic foot or per barrel.

26. *EEI Pocketbook,* p. 22.

27. FPC, *Steam Plant,* 1967, p. xiv.

28. These are hydroelectric facilities where, during the off-peak hours, water is pumped into a reservoir above the dam to again fall through the turbines, producing power when needed. The country's first pumped storage project was Connecticut Light & Power's 21 Mw Rocky River facility. It was in operation in 1929.

29. Many references are made in the industry literature to the production economies of scale present in electric power generation. The most complete analyses of these economies of scale include: Malcolm Galatin, *Economies of Scale and Technological Change in Thermal Power Generation;* Suilin Ling, *Economies of Scale in the Steam-Electric Power Generating Industry, An Analytical Approach;* Thomas G. Cowing, "Technical Change and Scale Economies in an Engineering Production Function: The Case of Steam Electric Power," *Journal of Industrial Economics;* David A. Huettner and John H. Landon, "Electric Utilities: Economies and Diseconomies of Scale" (unpublished); Charles E. Olson, *Cost Considerations for Efficient Electric Supply* (East Lansing: Division of Research, Graduate School of Business Administration, Michigan State University, 1970); and Marc Nerlove, "Returns to Scale in Electricity Supply," in *Measurement in Economics,* Carl Christ et al. (Stanford University Press, 1963), pp. 167-98. Nerlove concludes on pp. 186-87: "There is evidence of a marked degree of increasing returns at the firm level; but the degree of returns to scale varies inversely with output and is considerably less, especially for large firms, than that previously estimated for individual plants."

C. Maxwell Stanley, president of Stanley Consultants, has used data representative of January 1969 costs to estimate the changes in construction costs and heat rate as a function of unit size for coal-fired and nuclear generating units. His results follow.

Coal-Fired Generating Units
Comparative Costs and Heat Rates

Unit size (Mw)	Constr. cost ($/kw)	Heat rate (BTU/kwh)
290	174	8960
580	147 (15.5)*	8770 (2.12)
870	138 (6.5)	8600 (1.91)
1160	135 (2.2)	8500 (1.16)
2320	133 (1.5)	8430 (.83)

SOURCE: C. Maxwell Stanley, "The Impact of Changing Economics on Electric Utilities," *American Power Conference* (1969), table III, p. 12.
NOTE: Construction costs and heat rates based on first unit, 3,500/1,000/ 1,000 psi coal-fired, total enclosure, natural water-cooled unit.

* Figures in parentheses represent percentage reduction from the preceding unit size.

There are greater production economies in nuclear generation. Installed cost ($/kw) drops more for the same increase in unit size for a nuclear unit than for a fossil-fueled unit. The following table illustrates construction cost and fuel cost changes for selected nuclear unit sizes.

Nuclear Generating Units
Comparative Costs

Unit size (Mw)	Constr. cost ($/kw)	Fuel cost (mills/kwh)
580	195	1.55
870	165 (15.4)*	1.48 (4.5)
1160	150 (9.1)	1.44 (2.7)
2320	140 (7.2)	1.41 (2.1)

SOURCE: C. Maxwell Stanley, "Impact of Changing Economics," table IV, p. 13.

NOTE: Costs based on boiling water reactor technology with 2, 3, 4, and 8 loops, 70 percent load factor, 7 percent interest rate, December 1968 pricing of fuel and processing, levelized over the first ten years of the unit's life.

* Figures in parentheses represent percentage reduction from the preceding unit size.

To date there has been no known empirical examination of the size of the overall social costs related to electric power generation. Studies examining the amount of pollution per unit of output from different-sized electric power plants should be undertaken to determine the full costs of electric power generation.

30. The improvements in a unit's net heat rate diminish as operating pressures increase. Improvement in net heat rates rises at first and then begins to diminish as operating temperatures are further raised. These results are summarized as follows:

Change in operating conditions:

	Percentage improvement in net heat rate
I. *Increase in pressure*	
From 1,800 to 2,400 initial psig	2.0
From 2,400 to 3,500 initial psig	1.5 to 1.8

II. *Increase in temperature*

From 950°F to 1,000°F initial temperature	.6
From 1,000°F to 1,050°F initial temperature	.7
From 1,000°F to 1,050°F reheat temperature	.8
Addition of 2d 1,000°F reheat vs. 1,000°F initial, 1,000°F single reheat	1.5
Increase in 1st reheat from 1,000°F to 1,025°F and in 2d reheat from 1,000°F to 1,050°F	.8

SOURCE: D.W.R. Morgan, Jr., "Trends in Thermal Power" (Paper presented at Pacific Northwest Trade Association, 52d Conference, 14-16 September 1964), pp. 5-2 to 5-3.

Diminishing returns also can be observed from the reductions in construction and fuel costs and in heat rates as unit sizes are increased. See note 29, supra.

31. FPC, *Statistics, Privately Owned, 1970*, p. 763.
32. Total electric operating revenues include total revenues from sales of electricity plus other operating revenues, which include sales of water and water power, rent from electric property, sale of steam, wheeling charges (revenue from transmission of electricity generated by other utilities and transmitted over the company's facilities), and miscellaneous service revenue.
33. National Science Foundation, *Research and Development in Industry, 1969* (Washington, D.C.: Government Printing Office, 1970), p. 15. By 1973 the utilities had increased their commitment to R&D and reported spending over $239 million, representing .82 percent of total electric operating revenues.
34. William G. Meese, "Reasearch—Path to Progress in Electric Power," *American Power Conference* (1962), p. 120.
35. The exchange of power on a noncontractual basis is termed an interconnection. When this exchange is conducted on a contractual basis the organization is called a power pool.
36. Robert P. Liversidge, "Operating Experience with Large Generating Units," *American Power Conference* (1953), p. 385.
37. One of the findings of L.K. Kirchmayer, A.G. Mellor, J.F. O'Mara, and J.R. Stevenson in their article, "An Investigation of the Economic Size of Steam-Electric Generating Units," *American Institute of Electrical and Electronic Engineers, Transactions*, 74, pt. 3 (1955): 600, was that "the most economical pattern of system expansion is to add units between 10% and 7% of the size of the system studied."
38. See William M. Capron, ed., *Technological Change in Regulated Industries* for a broad discussion on the potential regulatory impact on technological advance.
39. William F. Crawford, "Our Changing Perspectives," *American Power Conference* (1961), p. 1.
40. Donald C. Cook, "Special Problems of Regulation" (Address before the Annual Regulatory Commission Development Short Course, Madison, Wisconsin, 28 August 1964), p. 7.
41. From correspondence with the writer, dated 19 July 1972.
42. See the *Economic Report of the President*, transmitted to the Congress February 1970 (Washington, D.C.: Government Printing Office, 1970), p. 107.
43. Harvey Averch and Leland L. Johnson, "Behavior of the Firm Under Regulatory Constraint," *American Economic Review* 52 (December 1962): 1052-69; also see Stanislaw H. Wellisz, "Regulation of Natural Gas Pipeline Companies: An Economic Analysis," *Journal of Political Economy* 71 (February 1963): 30-43; E.E. Zajac, "A Geometric Treatment of Averch-Johnson's Behavior of the Firm Model," *American Economic Review* 60 (March 1970): 17-25; Fred M. Westfield, "Regulation and Conspiracy,"

American Economic Reveiw 55 (June 1965): 424-43; Alvin K. Klevorick, "The Graduated Fair Return: A Regulatory Proposal," *American Economic Review* 56 (June 1966): 477-84; and William J. Baumol and Alvin K. Klevorick, "Input Choices and Rate-of-Return Regulation: An Overview of the Discussion," *Bell Journal of Economics and Management Science* 1 (Autumn 1970): 162-90.

44. According to Mr. Joseph C. Swidler, then chairman of the New York State Public Service Commission and former FPC chairman, there has never been a disallowance of R&D expenditures by a utility commission. (Information supplied from correspondence with the writer, dated 2 August 1972.)

45. Federal Power Commission, *Federal and State Jurisdiction and Regulation—Electric, Gas, and Telephone Utilities* (Washington, D.C.: Government Printing Office, 1967), table 12, p. 18.

46. *Federal Register* 35, no. 172 (3 September 1970): 13983-88.

47. The National Association of Regulatory Utility Commissioners' Uniform System of Accounts, used by twenty-five state public utility commissions in 1967, has been recently modified, incorporating substantially the FPC provisions for R&D.

48. Roy H. Dunham, "Growth of Steam Power on TVA System," *Power Engineering* 74 (July 1970): 38-39.

49. Leonard M. Olmsted, "10th Steam Station Design Survey," *Electrical World* 170 (21 October 1968): 83, states that power industry designers have decided more can be gained by exploiting scale economies than further improving the fuel cycle efficiency.

50. At times the manufacturers have had to guarantee the utilities alternative sources of power in order to persuade the utility to invest in new plant capacity employing new designs. Gulf General Atomic has promised the Public Service Company of Colorado gas turbine capacity if the Fort Saint Vrain gas-cooled nuclear power station is not operating by the guaranteed delivery date.

51. *Nucleonics* 22 (May 1964): 17.

CHAPTER III

1. Not everyone agrees with the widespread belief that the nation faces a power crisis. See " 'Sheer Nonsense' [EEI president A.H.] Aymond Declares to Suggestions of a Power Crisis," *EEI Bulletin* 38 (January 1970): 7.

2. The critical temperature and pressure of water is 705.4°F and 3,206.2 psi, where the properties of the saturated liquid and the saturated vapor are identical.

3. The most important recent change in power generation is probably the use of nuclear fission to heat water into steam to drive the turbine-generator; however, the nuclear reactor is merely replacing the conventional boiler as the source of steam. Another refinement in basic power technology occurred when alternating current was first used to transmit electric energy. This happened shortly after the Pearl Street Station was generating electric

power. A generating station in Buffalo, New York, began using an a.c. distribution system rather than the then prevalent direct current system.

4. See Appendix G for a representative list of utilities conducting group research in nuclear R&D.

5. "First Commercial Atomic Power," *Electrical World* 144 (25 July 1955): 74. The first electric power developed by a nuclear reactor was generated on 20 December 1951 at the AEC's reactor testing facility in Arco, Idaho. The reactor was an experimental breeder reactor (from *Nucleonics* 10 [February 1952]: 72).

6. A breeder reactor, as differentiated from a thermal reactor, produces, or "breeds," more fuel than it consumes. Without becoming too technical, the original fuel in the core of the reactor, Uranium 235, emits neutrons from the chain reaction at high speed. These fast neutrons are absorbed by the unfissionable Uranium 238 (U-238) surrounding the fuel. The U-238, after absorbing these "free" neutrons, gradually is transmuted into fissionable plutonium. This created plutonium can then be removed and used to fuel another reactor. The breeder reactor creates more "free" neutrons than a thermal reactor, producing more fuel than it consumes.

7. The reactor's start-up date was in December 1957.

8. Dan Braymer, "5th Nuclear Power Report," *Electrical World* 153 (16 May 1960): 63-82. This report provided information for the preceding summary of the AEC Power Demonstration Reactor Program.

9. An effective means of reducing particulate content from stack gases is to use electrostatic precipitators. The first such precipitator was installed and operated long before the government enacted pollution legislation. The first electrostatic precipitator was operating in 1924 at Detroit Edison's Trenton Channel Station.

10. *Electrical World* 172 (3 November 1969): 19.

11. A list of the major domestic power equipment manufacturers is presented in Appendix H.

12. The writer, perhaps naïvely, assumes that the conduct made public in the electrical conspiracy price-fixing case was an aberration from usual industry practice.

13. Not all electric utilities purchase equipment from foreign producers. Some companies, usually smaller utilities, subscribe to "Buy America" policies.

14. "The Utility Role in R&D . . .," *Electrical World* 161 (23 March 1964): 132.

15. Ibid., p. 130. Much of the engineering design work needed for constructing and building a utility power plant is performed outside the firm by consultants.

16. Research projects taken from William G. Meese, "Research—Path to Progress in Electric Power," *American Power Conference* (1962), pp. 120-21.

17. "The Utility Role in R&D . . .," p. 130.

18. The group of utilities is Central Illinois Light Company, Illinois Power Company, Dayton Power and Light Company, Indianapolis Power & Light Company, Kansas Power and Light Company, Louisville Gas and Electric

Company, Union Electric Company, and the following subsidiary companies of American Electric Power Company: Appalachian Power Company, Indiana & Michigan Electric Company, and Ohio Power Company. American Electric Power Service Corporation acts for the group in carrying out the joint research with AVCO Corporation. The EEI also supports this research.

19. Central Power and Light Company, Community Public Service Company, Dallas Power & Light Company, El Paso Electric Company, Gulf States Utilities Company, Houston Lighting & Power Company, Southwestern Public Service Company, Texas Electric Service Company, Texas Power & Light Company, and West Texas Utilities Company.

20. The members of this company and the above-mentioned nuclear utility groups are listed in Appendix G.

21. D. Bruce Mansfield, "American Power Industry Is Called Most Efficient, Reliable in the World," *EEI Bulletin* 38 (July-August 1970): 202.

22. Willis Gale, "The EEI's Expanded Research Program," *EEI Bulletin* 29 (June 1961): 185.

23. Walter H. Sammis, "Research and Progress," *EEI Bulletin* 32 (July 1964): 165.

24. Ibid.

25. "EEI Expands Its Research Program," *EEI Bulletin* 37 (March 1969): 87.

26. Information about the EEI Research Division was provided by John Endres. The $119.5 million figure represents the 1974 goal of the EPRI program sought from the privately owned utilities.

27. "The Utility Role in R&D . . .," p. 130.

28. Herbert I. Blinder, "Expanded 'R&D' Needed to Meet 'Energy Crisis,'" *Public Power*, July-August 1971, p. 9.

29. "The Utility Role in R&D . . .," p. 130.

30. Blinder, "Expanded 'R&D' Needed," p. 9.

31. Edwin H. Snyder, "The Role of the University in R&D for the Electric Utility Industry," *American Power Conference* (1968), p. 87.

32. Steam tables contain information about the physical properties of the vapor at various temperatures and pressures.

33. Meese, "Research—Path to Progress," p. 123.

34. The comments of Professor Paul C. Krause of Purdue University's Department of Electrical Engineering provided information about university-conducted power R&D.

35. "The Utility Role in R&D . . .," p. 135. Many of the smaller utilities do not seek engineers with advanced degrees. These companies may think that students with master's or doctoral degrees would not be sufficiently challenged by the "nuts and bolts" problems usually facing a utility. In 1969 it was reported by Senator Lee Metcalf that throughout the nation electric utilities employed a total of only eight Ph.D. engineers. (116 pt. 2, *Congressional Record*, 28 January 1970, p. 1701.)

36. Snyder, "The Role of the University," p. 87.

37. Meese, "Research—Path to Progress," p. 119.

38. *Statistics, Privately Owned*, 1970, p. xxxix.

39. Meese, "Research—Path to Progress," p. 120.

40. *Statistics, Privately Owned,* 1970, p. xviii and p. 766, respectively.
41. Such hostility between the private and public sectors of the utility industry may have diminished lately in intensity, but is still ever present.
42. "The Utility Role in R&D . . .," p. 132.
43. The preceding information was given to the writer in an interview with Mr. Dwon.
44. R.S. Talton, "Build First Automatic Coal-Fired Unit," *Electrical World* 156 (21 August 1961): 41-42.
45. When the infamous East Coast Blackout occurred on 9 November 1965, full-scale congressional hearings were called to determine the cause of the power breakdown.
46. A.H. Aymond, "Electric Energy Serves the Seventies," *EEI Bulletin* 38 (July-August 1970): 208.
47. Meese, "Research—Path to Progress," p. 120.
48. J. Grebe, "Why Is Dow-Detroit Edison Working on a Fast Breeder Reactor for Power?" *Nucleonics* 12 (February 1954) :13.
49. The Armor Research Foundation, affiliated with the Illinois Institute of Technology, performs various engineering and power research projects. Miller said that Insull, wanting to beat Consolidated Edison in being the first electric utility to use oil-insulated distribution cable, devoted many of Commonwealth Edison's resources to the task. Commonwealth Edison then did become the first utility to use such cable.
50. This information was given to the writer in a letter from Mr. Sporn dated 19 July 1972.

CHAPTER IV

1. Appendix D contains a sample of power generation innovations occurring before 1950.
2. The examiners of the list of innovations were asked to identify the ten most technologically significant. Two of the examiners, Professor El-Abiad and Mr. O'Connor, did so. Because their opinions differed as to which ten innovations were the most significant, all of the innovations that either man so judged are described.
3. The bracketed numbers refer to the number of the innovation as listed in Table 4.1.
4. Charles R. Earle, "Science Gives Us More Ideas, New Materials, for Still Greater Power Field Development," *Power Engineering* 62 (January 1958) :58.
5. "New 4500 psi Generating Unit Breaks Through Critical Steam Pressure Barrier," *Power Engineering* 57 (July 1953) :68.
6. Wilbur H. Armacost, "Research in the Field of Steam Generation," *American Power Conference* (1955), p. 131.
7. "The Retreat from 1,100°F Takes Effect," *Electrical World* 164 (25, October 1965) :5.
8. The Tait Station Unit 4 was in service before a similar boiler was operating at Metropolitan Edison Company's Portland Power Station.

9. Harry T. Akers, "Present Trends in Design of Large Turbine-Generators," *American Power Conference* (1970), p. 955.

10. Information given in a letter to the writer dated 24 April 1972.

11. As generators have continued to grow in capacity, the utilities and manufacturers have developed more efficient cooling systems for the generator rotor and stator. In the late 1950s oil-cooled and water-cooled generators were first commercially used by the electric utilities (see Appendix C). Recently, Westinghouse Electric Corporation and Massachusetts Institute of Technology announced the construction of experimental generators cooled with liquid helium to cryogenic temperatures. (See "Superconductivity Means More Megawatts," *Business Week*, 13 May 1972, pp. 150-54.)

12. Louis Gee, "A New Power Cycle Combines Gas Turbines with Steam Turbines," *American Power Conference* (1954), p. 229.

13. "Control Breakthrough at Little Gypsy," *Power* 103 (May 1959): 64.

14. Figures are taken from "Utility Engineers Discuss Research and Automation," *Power Engineering* 66 (December 1962) :67.

15. During seasonal periods when there is abundant water in the hydro units' reservoirs, utilities often base-load their hydro facilities and cycle those fossil-fueled units that have been designed for such operation.

16. Carl E. Bagge, "The Electric Power Industry in Crisis and Transition: The Need for a National Energy-Environment Policy," *American Power Conference* (1970), p. 12.

17. Ibid., p. 6.

18. Shortly after Kyger Creek Station began operation, American Electric Power Company's operating division, Indiana & Michigan Electric Company, operated its Clifty Creek Station with three 682 foot stacks.

19. G.W. Bice and A.F. Aschoff, "Status of SO_2 Removal Systems," *American Power Conference* (1969), p. 530.

20. Ibid., p. 534.

21. R.F. Roth and T.F. Steel, "Design and Selection of Hyperbolic Cooling Towers," 31 *Combustion* (January 1960) :42.

22. Ibid.

23. See note 29, chapter II.

24. Figures taken from Ebasco Services, Inc., Research Department, *1974 Business and Economic Charts*, New York (1974), p. 18. As of October 1974 there were: 51 operating nuclear plants having a total capacity of 33,760 Mw; 169 plants under construction or design with an expected capacity of 181,876 Mw; and 31 plants committed, under inquiry, or being planned, having an estimated capacity of 35,100 Mw.

25. D.W.R. Morgan, Jr., "Trends in Thermal Power" (Paper presented at the 52d Conference of the Pacific Northwest Trade Association (14-16 September 1964), table H.

26. The bidders for the AEC contract were: (1) Cisco Construction Company, Portland, Oregon; (2) Connecticut Light and Power Company; (3) Consumers Public Power District, Columbus, Nebraska; (4) Eastern Utilities Associates, Boston, Massachusetts; (5) Middle South Utilities, Inc. and Ebasco Services, Inc; (6) Nuclear Power Group, then composed of Pacific Gas and Electric Company, and Betchel Corporation; (7) Philadelphia Elec-

tric Company and Pennsylvania Power & Light Company; (8) South Carolina Electric & Gas Company; and (9) Dusquesne Light Company.
27. W.L. Felsen, "6th Report on Nuclear Power," *Electrical World* 155 (22
May 1961) :79.
28. In a PWR the light water, serving as the reactor's moderator and
coolant, is maintained under pressure to prevent it from turning to steam inside the reactor (at Shippingport the coolant was kept at 1,800 psi and
500°F). The water coolant is pumped through the reactor at high pressure,
absorbing the heat friction from the nuclear chain reaction. The heated coolant then leaves the reactor where it gives up thermal energy at a heat exchanger. The coolant then returns to the reactor core to be reheated. An in-
dependent (secondary) water cycle is used in PWRs. This water is heated
by the reactor coolant at the heat exchanger, generating steam for the turbine-generators. In contrast, a BWR produces steam inside the reactor,
eliminating the need for a heat exchanger and another water cycle. Unlike a
PWR, the steam produced by a BWR carries small amounts of short-lived
radioactivity.
29. Fensen, "6th Report on Nuclear Power," p. 79.
30. T.G. LeClair and C.L. Richard, "Recent Advances in High-Temperature
Gas-Cooled Reactors," *American Power Conference* (1965), p. 233.
31. See Appendix G for membership list.
32. F.C. Olds, "Gas Cooled Reactor's Standout Performance," *Power Engineering* 73 (October 1969) :42.
33. See note 6, chapter III for a simple description of the operation of a
breeder reactor.
34. See "Basic Problems in Central Station Nuclear Power," *Nucleonics* 10
(September 1952) :8.
35. See Appendix G for membership list.
36. "APDA History and Objectives," pamphlet published by The Detroit
Edison Company, p. 6.
37. Dan Braymer, "5th Nuclear Power Report," *Electrical World* 153 (16
May 1960) :73.
38. Adolph J. Ackerman, "Atomic Power Plants—What's Wrong with
Them?" *American Power Conference* (1965), p. 745. The figures do not include an estimated $4.5 million in AEC research assistance or $5 million in
waivers of AEC fuel charges.
39. Ibid.
40. Ibid., p. 746 and the *Wall Street Journal*, 30 November 1972, p. 14.
41. In January 1972 an announcement was made that TVA and Commonwealth Edison would join in a project to construct a much larger (500 Mw)
fast breeder reactor power station that was estimated to cost approximately
$700 million. Of this amount, $250 million would be paid by the electric
utility industry and $450 million by the AEC, which is financing the plant
as an R&D project. The $700 million figure is almost $200 million more than
the estimated cost of the project given in September 1971. By January 1975,
this demonstration plant was estimated to cost $1.74 billion and was planned
for a 1982 rather than 1980 completion date. (See the *Wall Street Journal*,
27 January 1975, p. 13.)

42. P. Dragoumis, S.J. Weems, and W.C. Lyman, "Ice Condensor Reactor Containment System," *American Power Conference* (1969), p. 347.

CHAPTER V

1. The hypothesis that the innovation ratios of each class are equal was accepted after statistical testing. The hypothesis was accepted at the .01 confidence limit.

2. To test whether the innovation ratios of other utility groups were significantly different than the ratio of the four largest utility systems, Group 1, the standard t-ratio was used. The calculated t-ratios for testing whether the innovation ratio of Group 1 was significantly different from that of Group 3, Group 5, Group 9, and Group 12 were, respectively: 1.25, 0.96, 0.75, and 0.76, none of which is greater than the critical value of t at the .05 confidence limit. Separate tests also revealed that the innovation ratios of Groups 2 and 12 were not significantly different ($t = 1.18$).

3. See Edwin Mansfield, *Industrial Research and Technological Innovation*, p. 91.

4. Appendix J identifies these electric utility systems and their affiliated operating utilities.

5. The method employed to weight the innovations is explained on page 89.

6. Mansfield's attempts at quantifying the effect of management background and education are not terribly encouraging; see Mansfield, *Industrial Research and Technological Innovation*, pp. 165-66.

7. See F.M. Fisher and P. Temin, "Returns to Scale in Research and Development: What Does the Schumpeterian Hypothesis Imply?" *Journal of Political Economy* 81 (January/February 1973) :56-70.

8. James Tobin, "Estimation of Relationships for Limited Dependent Variables," *Economica* 26 (January 1958) :24-36.

9. John W. Wilson, in his doctoral dissertation, "Residential and Industrial Demand for Electricity: An Empirical Analysis" (Cornell University, 1969), determined that price elasticity of industrial demand for electricity ranged from -1.65 to -2.40. The specific elasticity coefficient depended in large part on the type of production process employed and other related factors.

10. C.B. Broyden, "Quasi-Newton Methods and Their Application to Function Minimization," *Mathematics of Computation* 21 (1967) :368-81. The Full-Information Maximum Likelihood (FIML) method is a nonlinear estimating technique employed to determine the joint conditional distributions of two or more jointly dependent variables given the values of a set of predetermined variables. The estimation technique determines that set of parameters (all parameters are regarded as random variables) such that the value of one likelihood function of the data, given those parameters, is greater than the value of the likelihood function given any other set of parameters for a given model's specification. The estimates are obtained by maximizing the likelihood function, in logarithmic form, subject to all the *a priori* restrictions imposed by the structural equations of the model. The FIML estimators are consistent, asymptotically efficient, and asymptotically

normally distributed. The structural disturbances are assumed to be normally distributed.

11. Ohio Valley Electric Corporation reported no R&D expenditures in 1970.

12. See Joseph Schumpeter, *Business Cycles* (New York: McGraw-Hill, Inc., 1939); Joseph Schumpeter, *Capitalism, Socialism, and Democracy* (New York: Harper & Row, 1947); John K. Galbraith, *American Capitalism* (Boston: Houghton Mifflin Company, 1952); and H. Villard, "Competition, Oligopoly, and Research," *Journal of Political Economy*, December 1958, pp. 483-97.

13. The hypothesis that $\delta_1 \neq 1$ was rejected at the .05 confidence limit for each year.

14. See Edwin Mansfield, *Industrial Research and Technological Innovation*, pp. 39-40.

15. G.E.P. Box and D.R. Cox, "An Analysis of Transformations," *Journal of the Royal Statistical Society, Series B* 24 (1964) :211-43.

16. Mansfield, *Industrial Research and Technological Innovation*, p. 98.

APPENDIX A

1. This examination relies heavily on Martin G. Glaeser, *Outlines of Public Utility Economics*, Eli W. Clemens, *Economics and Public Utilities* (New York: Appleton-Century-Crofts, Inc., 1950), and Alfred E. Kahn, *Economics of Regulation, Principles and Institutions*, vol. I–II (New York: John Wiley & Sons, 1970-71).

2. Kahn, *Economics of Regulation*, II:xi.

3. See Paul MacAvoy, "Effectiveness of the Federal Power Commission," *Bell Journal of Economics and Management Science* 1 (Autumn 1970) :271-303; George J. Stigler and Claire Friedland, "What Can Regulators Regulate: The Case of Electricity?" *Journal of Law and Economics* 5 (1962) : 1-16; Roger C. Cramton, "The Effectiveness of Economic Regulation: A Legal View," *American Economic Review* 54 (May 1964) : 182-91; Harry Trebing, "What's Wrong with Commission Regulation?" pt I, *Public Utilities Fortnightly* 65 (12 May 1960) :660-70, pt II, *Public Utilities Fortnightly* 65 (26 May 1960) :738-50; and Stephen G. Breyer and Paul MacAvoy, *Energy Regulation by the Federal Power Commission* (Washington, D.C.: The Brookings Institution, 1974).

APPENDIX B

1. Much of this analysis is based on Andrew W. Kramer's *Power Plant Primer* (Barrington, Illinois: Technical Publishing Co., 1954).

2. Pulverized coal first was used in steam generation in 1918. See Appendix D.

3. Modern steam turbines may have more than one shaft (these multishaft units are called cross-compound machines). For simplicity, a single shaft machine is described in the example.

4. Information about turbine blade length was provided in a letter sent to the writer by Lloyd F. Kramer of the General Electric Corp.

APPENDIX E

1. John I. Yellott, "A Power Engineer Looks at Magnetohydrodynamics," *Power Engineering* 64 (April 1960) :72-75, *passim;* and Philip Sporn and Arthur Kantrowitz, "Magnetohydrodynamics—Future Power Process?" *Power* 103 (November 1959) :63-64.
2. Sporn and Kantrowitz, "Magnetohydrodynamics," p. 63.
3. Ibid.
4. Seymour Baron, "Are We Overlooking the Fuel Cell?" *Electrical World* 174 (1 December 1970) :44-46, *passim.*

Glossary
of
Abbreviations

a.c.   alternating current

BTU   British Thermal Unit

d.c.   direct current

F   temperature in degrees Fahrenheit

kw   kilowatt

kwh   kilowatt-hour

Mw   Megawatt; 1 Mw = 1,000 kw

Mwe   Megawatt electric

psi   pressure in pounds per square inch

psia   pressure in pounds per square inch absolute

psig   pressure in pounds per square inch gauge

rpm   revolutions per minute

Selected Bibliography

Books:

Alderson, Wroe; Terpstra, Vern; and Shapiro, Stanley J., eds. *Patents and Progress.* Homewood: Richard D. Irwin, Inc., 1965.

Capron, William M., ed. *Technological Change in Regulated Industries.* Washington, D.C.: The Brookings Institution, 1971.

Carter, C., and Williams, B. *Industry and Technical Progress.* New York: Oxford University Press, 1957.

Clemens, Eli Winston. *Economics and Public Utilities.* New York: Appleton-Century-Crofts, Inc., 1950.

Galatin, Malcolm. *Economies of Scale and Technological Change in Thermal Power Generation.* Amsterdam: North Holland Publishing Co., 1968.

Garfield, Paul J., and Lovejoy, Wallace F. *Public Utility Economics.* Englewood Cliffs: Prentice-Hall, Inc., 1964.

Jewkes, John; Sawers, David; and Stillerman, Richard. *The Sources of Invention.* 2d ed. London: Macmillan & Co., Ltd., 1969.

Kahn, Alfred E. *The Economics of Regulation: Principles and Institutions.* Vols. I-II. New York: John Wiley & Sons, Inc., 1970-71.

Ling, Suilin. *Economies of Scale in the Steam-Electric Power Generating Industry, An Analytical Approach.* Amsterdam: North-Holland Publishing Co., 1964.

Mansfield, Edwin. *Industrial Research and Technological Innovation.* New York: W.W. Norton & Co., Inc., 1968.

————, ed. *Monopoly Power and Economic Performance, The Problem of Industrial Concentration.* Rev. ed. New York: W.W. Norton & Co., Inc., 1968.

————. *The Economics of Technological Change.* New York: W.W. Norton & Co., Inc., 1968.

Nelson, Richard R.; Peck, Merton J.; and Kalechek, Edward D. *Technology, Economic Growth and Public Policy.* Washington, D.C.: The Brookings Institution, 1967.

Sporn, Philip. *Technology, Engineering, and Economics.* Cambridge, Mass.: The MIT Press, 1969.

151

————. *The Social Organization of Electric Power Supply in Modern Societies.* Cambridge, Mass.: The MIT Press, 1971.

Trebing, Harry M., ed. *Performance Under Regulation.* East Lansing: MSU Public Utilities Studies, Division of Research, Graduate School of Business Administration, Michigan State University, 1968.

Government Publications:

Energy R&D and National Progress. Prepared for the Interdepartmental Energy Study by the Energy Study Group, under the direction of Ali Bulent Cambel. Washington, D.C.: Government Printing Office, 1964.

Federal Power Commission. *National Power Survey, 1964.* Washington, D.C.: Government Printing Office, 1964.

————. *National Power Survey, 1970.* Washington, D.C.: Government Printing Office, 1970.

————. *Statistics of Privately Owned Electric Utilities in the United States.* Washington, D.C.: Government Printing Office, yearly.

————. *Statistics of Publicly Owned Electric Utilities in the United States.* Washington, D.C.: Government Printing Office, yearly.

————. *Steam-Electric Plant Construction Cost and Annual Production Expenses.* Washington, D.C.: Government Printing Office, yearly.

Articles:

Ackerman, Adolph. "Atomic Power Plants—What's Wrong with Them?" *American Power Conference,* 1965.

Adams, Francis. "Present Status of Power Pools." *Power Engineering,* October 1961.

Akers, Harry. "Present Trends in Design of Large Turbine-Generators." *American Power Conference,* 1970.

Armacost, Wilbur. "Research in the Field of Steam Generation." *American Power Conference,* 1955.

Bagge, Carl E. "The Electric Power Industry in Crisis and Transition: The Need for a National Energy-Environment Policy." *American Power Conference,* 1970.

Bice, G., and Aschoff, A. "Status of SO_2 Removal Systems." *American Power Conference,* 1969.

Brown, Seymour. "Are We Overlooking the Fuel Cell?" *Electrical World,* 1 December 1970.

"Control Breakthrough at Little Gypsy." *Power,* May 1959.

Cook, Donald C. "Special Problems of Regulation." An address given at the Annual Regulatory Commission Development Short Course, Madison, Wisconsin, 28 August 1964.

Crawford, William F. "Our Changing Perspectives." *American Power Conference,* 1961.

Dauber, Clarence. "Pipeline Transportation of Coal." *American Power Conference,* 1957.

Davis, W. Kenneth. "The Role of Government in Nuclear Power Development." *American Power Conference,* 1956.

Dickey, Paul. "Trends in Combustion and Steam Temperature Control." *American Power Conference,* 1953.

Dunham, Roy. "Growth of Steam Power on TVA System." *Power Engineering,* July 1970.

Ebdon, H. "The Trend Toward Reheat." *American Power Conference,* 1951.

Enos, John L. "Invention and Innovation in the Petroleum Refining Industry." In *The Rate and Direction of Inventive Activity.* Princeton: National Bureau of Economic Research, 1962.

Gaffert, Gustav. "Recent Improvements in Central Stations." *American Power Conference,* 1950.

Hamberg, D. "Size of Firm, Oligopoly and Research: The Evidence." *Canadian Journal of Economics and Political Science,* February 1964.

Hobson, J., and Lewis, W. "Research and the Electric Power Industry." *American Power Conference,* 1954.

King, Allen. "The Challenge of the Future in Electric Power." *American Power Conference,* 1960.

Kirchmayer, L.K.; Mellor, A.G.; O'Mara, J.F.; and Stevenson, J.R. "An Investigation of the Economic Size of Steam-Electric Generating Units." *American Institute of Electrical and Electronic Engineers,* Transactions, vol. 74, pt 3, 1955.

Kramer, Andrew. "Advancement of Nuclear Power, 1942-1962." *Power Engineering,* January 1963.

―――. "1955—Year One of Commercial Nuclear Power." *Power Engineering,* January 1956.

Kuznets, Simon. "Inventive Activity: Problems of Definition and Measurement." In *The Rate and Direction of Inventive Activity.* Princeton: National Bureau of Economic Research, 1962.

LeClair, T., and Pickard, C. "Recent Advances in High-Temperature, Gas-Cooled Reactors." *American Power Conference,* 1965.

Leitz, F.; Imhoff, D.; and McEwen, L. "Technological Developments of Nuclear Superheat." *American Power Conference,* 1966.

Macmillian, John. "Nuclear Power and Supercritical Steam Cycles." *American Power Conference,* 1965.

Mansfield, Edwin. "Innovation and Technical Change in the Railroad Industry." In *Transportation Economics,* a Conference of the Universities— National Bureau Committee for Economic Research, 1965.

Meese, William. "Research—Path to Progress in Electric Power." *American Power Conference,* 1962.

Miller, K. "Opportunities for Cooperative Research in the Electric Utility Industry." *American Power Conference,* 1957.

Minasian, Jora A. "The Economics of Research and Development." In *The Rate and Direction of Inventive Activity.* Princeton: National Bureau of Economic Research, 1962.

Morgan, D.W.R., Jr. "Trends in Thermal Power." A paper presented at the

52d Conference Pacific Northwest Trade Association, 14-16 September 1964.

"New 4500-psi Generating Unit Breaks Through Critical Steam Pressure Barrier." *Power Engineering*, July 1953.

Parker, E., and Baird, J. "Manufacturers' Research in Relation to Electrical Utilities." *Electrical Engineering*, March 1955.

Peck, Merton J. "Inventions in the Postwar American Aluminum Industry." In *The Rate and Direction of Inventive Activity*. Princeton: National Bureau of Economic Research, 1962.

Peterson, Robert. "Progress and Prospects in Electric Power." *American Power Conference*, 1966.

"Plant Design Advances Set New Records." *Power*, November 1954.

"Prestressed Concrete Pressure Vessels." *Nucleonics*, September 1965.

"Recent Advances in Reactor Technology." *Nucleonics*, May 1953.

Reti, George. "New Combined Cycle for Gas Turbines Offers High Efficiencies." *Power Engineering*, May 1965.

Rowand, W. "Supercritical-Pressure Steam Production Triples Capacity, Cuts Heat Rate 40%." *Electrical World*, 6 September 1954.

Sanders, Barkev S. "Some Difficulties in Measuring Inventive Activity." In *The Rate and Direction of Inventive Activity*. Princeton: National Bureau of Economic Research, 1962.

Scherer, F.M. "Size of Firm, Oligopoly and Research: A Comment." *Canadian Journal of Economics and Political Science*, May 1965.

―――. "Firm Size, Market Structure, Opportunity, and the Output of Patented Inventions." *American Economic Review*, December 1965.

Schmookler, Jacob. "Bigness, Fewness, and Research," *Journal of Political Economy*, December 1959.

―――. "Technological Change and Economic Theory." *American Economic Review*, Papers and Proceedings, May 1965.

Schroder, K., and Haack, P. "Steam Power Plant in the Final Phase of Present Day Metallurgy and Economy." *Combustion*, December 1967.

Seaborg, Glenn. "Nuclear Power—Status and Outlook." *American Power Conference*, 1970.

Seelye, Howard. "Research in the Electric Utility Industry." *American Power Conference*, 1955.

Skrotzki, B.G.A. "Superpressure Plants Lead Progress Parade." *Power*, October 1958.

Smith, B.A. "Technological Leadership in the Aluminum Industry." In *Technological Development and Economic Growth*. Edited by George W. Wilson. Bloomington: School of Business, Division of Research, Indiana University, 1971.

Smith, H. "Changing Generation Patterns." *Power Engineering*, November 1970.

Snyder, Edwin H. "The Role of the University in R&D for the Electric Utility Industry." *American Power Conference*, 1968.

Sporn, Philip, and Kantrowitz, Arthur. "Magnetohydrodynamics—Future Power Process?" *Power*, November 1959.

Stanley, C. Maxwell. "The Impact of Changing Economics on Electric Utilities." *American Power Conference*, 1969.

Swidler, Joseph C. "Research, Reliability, and the Environment in the Power Industry." Speech given at a meeting of the Institute of Electrical and Electronics Engineers, Inc., 2 February 1971.

————. "The Public Stake in Energy R&D." Speech given at the 40th Annual Convention of the Edison Electric Institute, 7 June 1972.

"The Utility Role in R&D . . ." *Electrical World*, 23 March 1964.

"U.S. Power Reactor Program." *Nucleonics*, July 1954.

Villard, Henry. "Competition, Oligopoly, and Research." *Journal of Political Economy*, December 1958.

Wallace, H., Jr. "Trends in Modern Central Station Boiler Design." *American Power Conference*, 1952.

Wessenauer, G.O. "The Story of TVA—Multipurpose Development." *Power Engineering*, February 1963.

Worley, James. "Industrial Research and the New Competition." *Journal of Political Economy*, April 1961.

Yellott, John I. "A Power Engineer Looks at Magnetohydrodynamics." *Power Engineering*, April 1960.